Translational Systems Sciences

Volume 31

Hiroshi Kawakami
Editor

Systems Design Based on the Benefits of Inconvenience

Springer

Editor
Hiroshi Kawakami
Faculty of Engineering
Kyoto University of Advanced Science
Ukyo, Kyoto, Japan

ISSN 2197-8832 ISSN 2197-8840 (electronic)
Translational Systems Sciences
ISBN 978-981-19-9587-3 ISBN 978-981-19-9588-0 (eBook)
https://doi.org/10.1007/978-981-19-9588-0

This Springer imprint is published by the registered company Springer Nature Singapore Pte Ltd.
The registered company address is: 152 Beach Road, #21-01/04 Gateway East, Singapore 189721, Singapore

Preface

This book is for innovators, designers, engineers, and researchers who are going to create new human-centered systems. The key concept of this book is Fuben-eki that is the Japanese term for "benefits of inconvenience (BoI)." Generally, human-machine systems are developed aiming to provide users convenience. In this case, beneficial aspects of inconvenience are ignored. On the other hand, all chapters of this book focus on BoI and describe commentaries and case studies in the field of product design, system design, and service design.

Originally, Fuben-eki emerged in the research field of mechanical engineering, especially focusing on human-machine systems. The concept soon spread around such fields of design studies as interaction design, interface design, and service design. Based on the notion of BoI, the products and services require users some effort and in turn give users some benefits that cannot be gained without inconveniences.

Funded by the Grant-in-Aid for Scientific Research of the Japan society for the promotion of science, the research project of BoI was organized. The society of instrument and control engineering, a Japanese engineering society, has published special issues of BoI in its journal twice in Japanese. The author of this book includes the member of the research project, and the chapters of this book are selected from the contributions of the special issues.

Kyoto, Japan

Hiroshi Kawakami

Contents

About the Editor

Hiroshi Kawakami is a professor at Kyoto University of Advanced Science, in Japan. He received his Bachelor's, Master's, and Dr. Eng. degrees from Kyoto University. He started his career at Okayama University as an assistant professor. He joined Kyoto University, where he was an associate professor of the Graduate School of Informatics and a program-specific professor of the Design Unit. His research interests include systems design, where he has proposed FUBEN-EKI that stands for designing systems based on appreciating "benefits of inconvenience." He received best paper awards of Transactions of the Society of Instrument and Control Engineering (1990, 2001, 2013), the Transactions of Human Interface Society (2009, 2018), and *Journal of Society of Automotive Engineers of Japan* (2014).

Chapter 1
Fuben-Eki: Japanese for "Benefit of Inconvenience" that Illustrates a New Value Orthogonal to Efficiency and Functionality

Hiroshi Kawakami

Abstract Fuben-eki is Japanese. Fuben means inconvenience and Eki means benefits. Fuben-eki means benefits that are gained only by inconvenience. Pursuing benefits employing not convenience but inconvenience sounds like one of the old Buddhist teachings, but in fact, it is relatively new. Providing Fuben-eki to users is one of the principles of system design, especially human–machine systems. This principle is effective in several design fields. They are not just confined to designing physical systems, e.g., product design but spread to service and business designs. As the general introduction of this book, this chapter introduced a definition of Fuben-eki (benefit of inconvenience: BoI) and discussed the common properties of design fields that are effective to introduce Fuben-eki. The property was elucidated as having a value axis orthogonal to efficiency and functionality. In the second half of this chapter, design fields were introduced. They contribute commentary and case studies to this book. The relationships between these fields and Fuben-eki were explained in short. They will help readers of this book to find an overlap between their interests and this book.

Keywords Benefit of inconvenience · System design · Human machine systems · Design thinking

1.1 Introduction

There are things and matters that are inconvenient in the sense that they require users and others involved to spend time and effort and use their brains, but that provide benefits because of that inconvenience. In general, inconvenience should be eliminated when it is said that "necessity is the mother of invention," and

---

H. Kawakami (✉)
Kyoto University of Advanced Science, Ukyo, Kyoto, Japan
e-mail: kawakami.hiroshi@kuas.ac.jp

H. Kawakami (ed.), *Systems Design Based on the Benefits of Inconvenience*, Translational Systems Sciences 31, https://doi.org/10.1007/978-981-19-9588-0_1

avoiding inconvenience can be a motive force for new technological development. On the other hand, there are such research areas as human–machine systems, human interfaces, and human-centered design, where interaction between humans and artificial systems is the essence. In these areas, human interaction with the artificial system should not be avoided as an inconvenience. Automated systems that eliminate human operations are no longer human–machine systems, and their human interfaces are no longer anything more than display devices. The operator is no longer the observer. This argument is valid not only in the design of physical systems, but also in the design of services, businesses, and policies. In the first place, the essence of services and business is that people are included in the system.

In 2012, "Journal of the Society of Instrument and Control Engineers (ISSN 0453-4662)" featured an issue titled "System Design based on Further Benefit of a Kind of Inconvenience" that attempts to actively utilize the benefits of inconvenience in a variety of fields in Japanese (Kawakami et al., 2012). Since then, the discussion of "benefits of inconvenience (BoI)" has not only spread to various research fields but also jumped over the boundaries of research. The same journal summarized the trends of "systems that give users benefits of inconvenience" in various fields in 2021 in Japanese (Kawakami et al., 2021).

1.2 Benefits of Inconvenience (BoI): Values Orthogonal to Efficiency and Functionality

This book is a collection of commentaries and case studies related to BoI. Because the term BoI is unfamiliar, in recent years it has sometimes been misused as "the counterpart of convenience." It is also sometimes mistakenly thought to be an old Buddhist term, but in fact, it is relatively new.

Each chapter of this book consists of studies in research areas where BoI is a keyword. Those areas are either directly related to the "Society of Instrument and Control Engineers (SICE)" (The Society of Instrument and Control Engineers, 2022) or are slightly related to SICE even though they may not usually be included in the scope of discussion. Authors of each chapter of this book examined the common denominator of all those studies and found that the research areas are concerning systems that have a value axis that is orthogonal to efficiency and functionality. We usually tend to think that it is good enough to endorse the improvement of efficiency and functionality. On the other hand, each chapter of this book presents a different (not contrary, but independent) axis of value.

From an engineering standpoint, BoI is considered when focusing on "systems that include human." Generally speaking, it is safe to assume that engineering is for the pursuit of convenience. In contrast, this book is oriented to actively let users experience inconvenience.

1.3 Rethinking Convenient and Affluent Society

The phrase "convenient and affluent society" sounds familiar, and even grating. It is a catchy and frequently used phrase for B to C, and as of 2021, a keyword search will find 150,000 hits on Japanese pages. However, when I look inside myself, who accepted this phrase without any thought, I realize that my deep psychology assumed the relationship shown in Fig. 1.1.

Let us consider what convenience/inconvenience is. It is a question that is difficult to answer when faced squarely. The answer to this question is attempted in other commentaries in this book, but for now, in this section, the answer is given in terms of objectively observable phenomena that can be treated in an engineering manner. Namely, "less effort required to accomplish a task" is called "convenience" in this chapter and "effort" here is defined as the following two types (Sawaragi, 2018).

- Physical effort: Often time-consuming, but not exclusively.
- Mental effort: Consuming cognitive resources (attention, memory, thinking, etc.)

In this way, at least when designing systems that include humans, it seems too cheap for the creators to think that all they have to do is provide convenience.

I conducted thought experiments with students and researchers. The following is a selection of the imaginary services obtained by the experiments.

- Assembling plastic model instead of you.
- Beating the last boss of computer games instead of you.
- Kneading "Nerunerunerune" instead of you.
- Carrying you to the top of "Mt. Fuji" by escalator.

"Nerunerunerune" is a famous Japanese snack, and kids enjoy kneading it into candy by themselves. "Mt. Fuji" is the highest mountain in Japan and many people enjoy climbing it.

All these services are superfluous to the user. Other examples, such as a drone that detects four-leaf clovers (Ozaki et al., 2018), also suggest the existence of tasks that lose their meaning when "hassle-free and convenient" is provided.

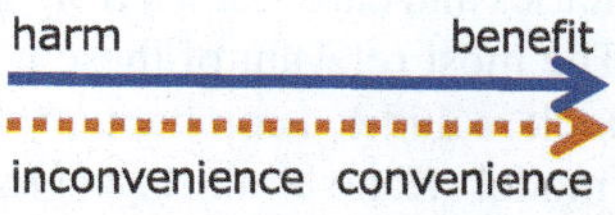

Fig. 1.1 The dependent relationship between convenience and benefit. Schematic representation of the assumptions behind the statement that convenience is always beneficial. Given that an axis representing "convenience/inconvenience" and an axis representing "benefit/harm" can be set up, these two axes can be considered to point in the same direction or to be identical. Note that this is only a schematic of a conceptual relationship. Strictly speaking, it is difficult to imagine that an interval or ratio scale can be introduced for "convenience/inconvenience" (i.e., an absolute value of "inconvenience" cannot be set), and it seems impossible to define an axis or a point where they intersect

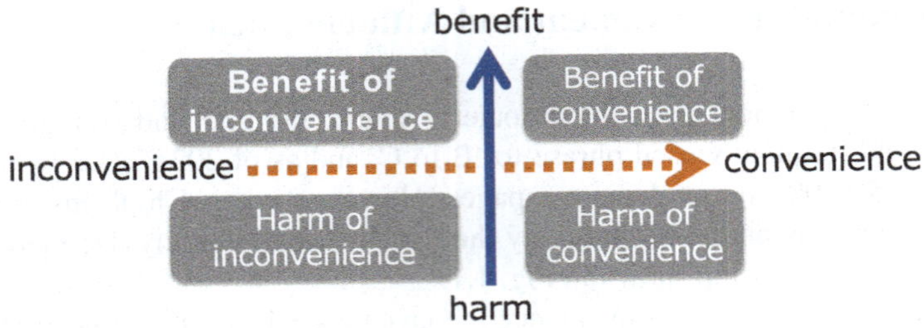

Fig. 1.2 The orthogonal relationship between convenience and benefit. Different from the relationship shown in Fig. 1.1, where convenience is always beneficial, this figure shows that convenience can be sometimes beneficial, and in some cases harmful. In this case, the axes representing "convenience/inconvenience" and "benefit/harm" are orthogonal, and four quadrants can be defined. Same as in Fig. 1.1, this figure only schematically represents a conceptual relationship, and the intersections of the axes do not represent such zeros as on a ratio scale

Once the existence of such tasks is acknowledged, the relationship between the two axes shown in Fig. 1.1 is altered as shown in Fig. 1.2.

Namely, convenience and benefit are considered to be independent and are represented schematically by orthogonal axes. Furthermore, the four quadrants defined by the two axes are named as shown in Fig. 1.2. We can plot the "events in which the convenience of saving time and effort causes harm," as described in the previous thought experiment examples, in the quadrant named "harm of convenience" (lower right space). The quadrants named "benefit of convenience" and "harm of inconvenience" (upper right and lower left) were originally placed on the left and right sides, respectively, of Fig. 1.1, and are nothing new. The remaining quadrant is the "benefit of inconvenience" (upper left). In this quadrant, events that provide users with experiences that are beneficial only because they are inconvenient are placed.

1.4 Benefit of Inconvenience in Manufacturing

This book features commentaries and case studies from areas that are closely related to and surrounding SICE. The most relevant of these is the area of manufacturing. Concerning this area, this book includes a commentary from the viewpoint of value engineering, and the Kano model (Kano et al., 1984) referred to therein provides a more general viewpoint of manufacturing. Therefore, with some overlap permitted, this chapter refers to it as well.

The Kano model classifies product quality based on the assumption that physical fulfillment and customer satisfaction are independent, as shown by the orthogonal axes in Fig. 1.3.

In the Kano model, "one-dimensional quality" is a relationship implicitly assumed by most people during the period of rapid economic growth, in which

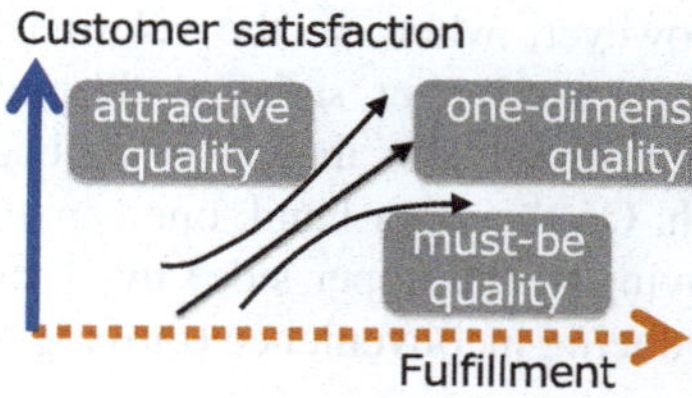

Fig. 1.3 Classification of product quality by the Kano model (ref. (Sawaguchi, 2020) partially abbreviated). The Kano model classifies product quality using a plane stretched on two axes: physical fulfillment and customer satisfaction. Three of them are named attractive quality, one-dimensional quality (performance quality), and must-be quality (basic quality), all of which are represented in the figure by monotonically increasing arrows

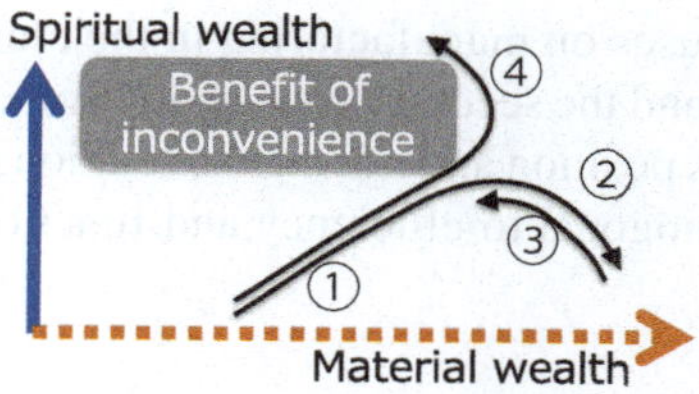

Fig. 1.4 Schematic explanation of BoI with the aid of the two axes of the Kano Model (Reference Sawaguchi (2020), p.33). This plane is defined by interpreted two axes in the Kano model. In this plane, a conceptual explanation of BoI is going to the upper left. A partial revision of this diagram will be included in the section on Value Engineering Perspective in this book

customer satisfaction increases almost linearly with improvement in the physical fulfillment of the product, and is represented by a straight line in Fig. 1.3. Other qualities include "attractiveness quality," which does not cause dissatisfaction even when physical fulfillment is low, and "must-be quality," which does not improve satisfaction much even when physical fulfillment increases. These three categories of quality are consistent with our intuition when we interpret "higher functionality" as an increase in physical fulfillment.

It is noteworthy that all three qualities are monotonically increasing (Δsatisfaction/Δfulfillment >0), and schematically they all rise steadily as shown in Fig. 1.3. Here, it can be seen that in the field of manufacturing, it is assumed that higher functionality will always satisfy customers (to varying degrees).

Let us interpret each axis in the Kano model as follows: the horizontal axis means material wealth (physical fulfillment: objective) and the vertical axis means spiritual wealth (customer satisfaction: subjective). This interpretation emphasizes the contrast between physically objective and spiritually subjective wealth and allows us to consider a plane defined by the axes as shown in Fig. 1.4.

The arrows labeled ① through ④ in the figure represent the following matters. First, ① when moving from a state, which is inconvenient at the beginning and is in itself a hindrance to life, to a more convenient and efficient state, material wealth and spiritual wealth have a monotonically increasing relationship (Δspiritual_wealth

/ Δmaterial_wealth >0). However, when ② the relationship breaks down, in other words, when the user is no longer satisfied despite physical fulfillment (Δspiritual_wealth / Δmaterial_wealth <0), then ③ going back in time may be one way to regain spiritual wealth. On the other hand, one can also be oriented toward gaining spiritual wealth (moving to the upper side) by ④ more actively reducing physical fulfillment and introducing inconvenience (moving to the left side).

1.5 Tracing Changes in Commentary in the Special Issue on the BoI

The previous chapter focuses on manufacturing in the close interest of the Society, which published the first and the second issues about BoI, and positioned BoI in that context. Starting from this position and tracing the relational networks, the fields that share the "value axis orthogonal to efficiency and functionality" spread in various directions.

1.5.1 Commentaries of BoI in 2012

"Journal of the society of instrument and control engineers" featured a special issue on "Systems that Utilize the Benefits of Inconvenience" in its August 2012 issue (the first issue in 2012) (Kawakami et al., 2012). The first issue consists of topics that are associated with the keyword "inconvenience" at that time, and that are on the periphery of systems engineering.

Among those topics, "the relationality-oriented systems design" (Shimohara, 2021) is explained. As is now well known, in many systems, people, objects, and matters are related (interdependent) and function in coordination and interlocking. Without realizing the perspective from which systems are viewed in this relationality, there was a trend in the past to view individual objects as closed systems, and to aim for higher functionality and efficiency of the system itself. This tendency, in turn, alienates the function of the entire system, which is composed of people, things, and matters. Considering this situation, the perspective of "viewing the system from the perspective of relationships" was reemphasized, and a system design based on this perspective was proposed.

One of the methods used is "subtraction." The method is that, although some functions seem to be subtracted when viewing an object alone, the subtraction conversely draws out the abilities and motivation of people, and harmonizes the whole that consists of people, objects, and matters. From the user's point of view, a single object looks (superficially) inconvenient. A study group on relational systems has been established in our society. Since the research trend is described in the materials of the study group, there is no contribution to this book.

In the first issue, we also received contributions on Shikakeology (Matsumura et al., 2015) that attempts to solve social problems through "mechanisms" (Shikake in Japanese) that change people's awareness and behavior. Even before the Nobel Prize in Economics was awarded to nudge theory in behavioral economics in 2017, attempts were being made to solve social problems in general, without being confined to economics, using mechanisms (including nudges) to induce human behavior. Even in recent years, we routinely see devices such as dolls placed on chairs in waiting rooms and footprint marks on the floor in front of supermarket checkout counters to induce COVID-19 infection prevention behavior.

By the way, we have all experienced that when we are told not to do something, we want to do it, and when we know something is hidden, we want to see it. People are sometimes motivated to do something induced by the inconvenience of being restricted. Here is the overlap of BoI and Shikakeology. Making something inconvenient is used to be a mechanism to motivate people to act. From the viewpoint of BoI, one of several categories of benefits is "motivation" and from the viewpoint of Shikakeology, one of several categories of mechanism is "making it inconvenient." This book has not received any contribution from Shikakeology. It has already gone beyond the academic framework and is in the phase of practical social implementation, which is well known to us without the need for contributions.

Other commentary included contributions from safety engineering, a field that has been studied extensively for some time, and media biotopes (Suto, 2011), an emerging field at the time. In a nutshell, the touchpoint of safety engineering and BoI is that "safety and security are incompatible." Risk-homeostasis (Wilde, 2001) is a well-known theory that homeostasis found in nature also manifests itself in the human psyche. We have all experienced the temptation to take risky actions when we feel safe, and to take safe actions when we feel insecure. In other words, although it is convenient for humans to improve the safety of a single object, it makes them feel secure and shifts their own behavior to the side of danger, and the safety of the human-machine system as a whole does not change (homeostasis), but rather the nature of danger transforms.

1.5.2 Commentaries of BoI in 2021

The above commentaries were contributed from different fields, but they essentially have something in common. They gave us viewpoints from which we recognize things and matters when we start a new discussion. Since then, these viewpoints have been used in a variety of ways, but the focus of this book is on their use in designing and producing new things. This is also consistent with engineering in the sense that the main concern is to create (design) new things that have never existed before. The commentary part of the special issue on "Journal of the society of instrument and control engineers" in 2021 (the second issue in 2021) (Kawakami et al., 2021) consists of the following contributions.

First, a set of the necessary conditions to be systems with BoI are examined from the standpoint of systems engineering. This set of conditions can be used to check the design results. Experiences that the design results give the user can be analyzed by the set of conditions whether the user's inconvenience yields the user's benefits. It can also be used in the form of designing a new system in the direction of satisfying the conditions without omission. This is like Lean Canvas (Maurya, 2012) for planning (designing) a business model.

Next, design in engineering is generally considered to be a field of technology. In this case, the design of products is implicitly oriented toward increasing functionality and reducing production costs. In this sense, designing something that forces users to experience inconvenience seems out of the question. On the contrary, if we broaden the perspective a little and look at the design of things that provide value to users, the design of things that provide value in the form of "inconvenience" is also an object of engineering. From the field of value engineering (VE), which has conventionally defined "value = function/cost" and provided on-site technology, a commentary is given that provides the framework for understanding inconvenience as a new value.

As described above, in engineering, the design target is often a thing (product). On the other hand, in academia, the design target is not confined to products but also includes businesses and services. In this book, the case of service design is explained as an example of the design of matters. Here, too, the idea that there is a value axis orthogonal to efficiency and functionality works well.

Design studies are known as an academic field that encompasses the design of things and the design of matters described above. Engineering and design share the aspect of conceiving and creating new things that have never existed before. In this sense, engineering and design studies should be adjacent. From an engineering perspective, design studies can be considered a peripheral area. However, perhaps because the design has not been recognized as an academic discipline for so long, the term "design studies" seems to be rarely seen in the societies of engineering in Japan. In contrast, this book explains the position of inconvenience in design studies for engineers.

In addition, ideation support was positioned as one of the application cases of BoI in the first issue, but in recent years when the term "innovation" has become overused, idea support has become an important topic of engineering interest. Therefore, the second issue adds a commentary on the position of BoI in ideation support.

1.6 The Transition of Case Studies in the First to Second Issues on BoI

In the first issue in 2012, the followings are introduced as examples of systems that take advantage of the benefits of inconvenience. Namely, An event called

"BiblioBattle" is an example of community design, a mobile terminal for sight-seeing navigation is an example of designing matters through designing things, an artificial hand and a stick-type metronome are examples of designing things, an automobile driving support system and a dysarthria support system are examples of interface design, a trash can robot is an example of relational system design, and extreme thinking is an example of idea support.

Weak robots (Okada, 2012) and tourism engineering were included in the second issue, following the first issue. The authors explain with concrete examples how they provide users with "value independent of efficiency and high functionality." In addition, the following case studies have been added to the second issue.

First, systems that support human activity by interacting with them is described with its relation to inconvenience. Although it may sound intuitively contradictory that obstructing can lead to support, each of the examples presented is certainly convincing.

Next, as an example of service design, a service of travel commerce inspired by BoI is presented. The app developed for this service has been downloaded as many as 150,000 times in the first year after its release.

Finally, as a concrete example of experience design that provides users with BoI, a chapter of this book presents works submitted to the student design competition sponsored by the "Japan Industrial Design Association (JIDA)" (Japan Industrial Design Association, 2022). JIDA is the only national organization of industrial designers in Japan, and BoI has been adopted as the theme of the Student Design Award organized by its Kansai block, with many students applying every year.

1.7 Conclusion

As a general introduction to this book, this chapter first defined the term "benefit of inconvenience." It also attempted to explain that related commentaries and examples commonly have a "value axis orthogonal to efficiency and functionality" and its image by several diagrams in which the two axes are orthogonal to each other. I hope these diagrams support readers to find some overlap between their interests and this book.

References

Japan Industrial Design Association (2022). https://www.jida.or.jp/, last accessed 2022/11/22.

Kano, N., et al. (1984). Attractive quality and must-be quality. *Quality, 14*(2), 147–156. https://doi.org/10.20684/quality.14.2_147

Kawakami, H., et al. (2012). Special issue: System design based on further benefit of a kind of inconvenience. *Journal of the Society of Instrument and Control Engineers*. (ISSN 0453-4662), *51*(8), 699–770.

Kawakami, H., et al. (2021). Special issue: Systems that give users benefits of inconvenience. *Journal of the Society of Instrument and Control Engineers*. (ISSN 0453-4662),, *60*(12), 831–879.

Matsumura, N., et al. (2015). Shikakeology: designing triggers for behavior change. *AI & SOCIETY, 30*, 419–429.

Maurya, A. (2012). *Running lean: Iterate from plan a to a plan that works*. O'Reilly Media.

Okada, M. (2012). *Weak robot (ISBN 978-4-260-01673-5)*. Igaku-Shoin Ltd.

Ozaki, M. Y., et al. (2018), Dest-logy—A drone in search for the four-leaf clover, https://www.youtube.com/watch?v=GFlyESz0wOg, last accessed 2022/9/25.

Sawaguchi, M. (eds.) (2020). Implementation of Benefits of Inconvenience (ISBN978-4-7649-6011-4). Kindai kagaku sha co., ltd., Tokyo, p.33.

Sawaragi, T. (Ed.). (2018). *Artifact design (ISBN 978-4-320-00602-7)* (p. 246). Kyoritsu Shuppan co., ltd..

Shimohara, K. (2021). Relationality Design and Relationality-oriented Systems Design, Advances in Knowledge-Based and Intelligent Information and Engineering Systems (DOI 10.3233/978-1-61499-105-2-1962), 1962-1971.

Suto, H. (2011). Media biotope: Media designing analogous with biotope. *International Journal of Computer Information Systems and Industrial Management Applications*. (ISSN 2150-7988), *3*, 264–270.

The Society of Instrument and Control Engineers (2022). https://www.sice.jp/english/, last accessed 2022/11/22.

Wilde, G. J. S. (2001). *Target Risk 2*. (ISBN 978-0969912439). Pde Pubns.

Hiroshi Kawakami is a professor at Kyoto University of Advanced Science, in Japan. He received his Bachelor, Master, and Dr. Eng. degrees from Kyoto University. He started his career at Okayama University as an assistant professor. He joined Kyoto University, where he was an associate professor of Graduate School of Informatics and a program-specific professor of Design Unit. His research interests include systems design, where he has proposed FUBEN-EKI that stands for designing systems based on appreciating "benefits of inconvenience." He received best paper awards of Transactions of the Society of Instrument and Control Engineering (1990, 2001, 2013), the Transactions of Human Interface Society (2009, 2018), and Journal of Society of Automotive Engineers of Japan (2014).

Chapter 2
Ideation Support for Designing Systems with Benefits of Inconvenience

Hiroshi Kawakami

Abstract For creating new products and services, one of the traditional design policies is thinking of it enough to provide convenience to users. On the other hand, a set of popular designs has revealed that the policy is no longer valid. Alternative policies implicitly exist. Among them, this chapter focused on "providing users benefits of inconvenience (BoI)." For designing new systems under such policy, this chapter summarized four methods of thinking derived from many years of design practice. The four types of methods can correspond to four transitions of the quadrants on the 2D plane that are defined by two axes, i.e., the axis of convenience/inconvenience and the axis of benefit/harm. The four types are named according to their characteristics, i.e., Value Mining type, Problem-solving type, Emergence type, and Innovation of Meaning type. This chapter also introduced two tools that support these types of thinking. The one is BoI cards that show how to make things inconvenient and which kind of benefits will be derived from those inconveniences. The other is the necessary conditions for judging systems that provide users BoI.

Keywords Benefit of Inconvenience · System Design · Design Thinking · Ideation Support

2.1 Introduction

In Japan, a machine for cooking rice, a staple food, is indispensable in most households. Rice cookers are convenient home appliances that cook rice on electricity, have a timer, automatically go into warming mode after cooking, require little time and effort of you, and can be used to cook rice to your liking simply by selecting a preset menu. Companies are competing to offer even higher functionality. On the

H. Kawakami (✉)
Kyoto University of Advanced Science, Ukyo, Kyoto, Japan
e-mail: kawakami.hiroshi@kuas.ac.jp

H. Kawakami (ed.), *Systems Design Based on the Benefits of Inconvenience*,
Translational Systems Sciences 31, https://doi.org/10.1007/978-981-19-9588-0_2

other hand, recipes for cooking rice on a gas stove and in an earthenware pot are also competing on the web. As of 2021, a search for "donabe gohan (Japanese for earthenware and cooked rice)" yields seven-million page hits. From the postings, it seems that they are not forced to endure inconvenient methods because earthenware pots are less expensive than rice cookers. On the contrary, there are many positive descriptions of the time and effort required to cook in earthenware pots.

There are healing home-use games in which you can make your avatar appear in a virtual space (e.g., a deserted island) and lead a heartwarming life. One of the software packages released in 2020 incorporates real-world "inconveniences" into the virtual space. For example, stores in the virtual space are only open during the daytime in the real world. This would be inconvenient for game play. However, the game was a worldwide hit.

Thus, when creating new goods and services, the traditional policy that it is enough to pursue convenience in a straightforward manner is no longer valid. It is suggested that an alternative policy exists (implicitly) among developers and designers. One such policy seems to be the use of "take advantage of benefits derived from inconvenience" in the above two examples.

The Japan Industrial Design Association (JIDA) (Japan Industrial Design Association, 2022) must have intuited the existence of this policy. The theme of the student design competition sponsored by a block of JIDA was "Benefit of Inconvenience (BoI)." In other words, professionals who are in the business of creating new things and asking the world about them are focusing on BoI as a concept that they want the students who will lead the next generation to know.

However, it is not easy to come up with new things or events that allow users to experience inconvenience that gives users benefits. Firstly, we are so accustomed to thinking in the direction of making things convenient that we cannot think of ways to make them inconvenient. As the saying goes, "necessity is the mother of invention," we tend to think in the direction of eliminating inconvenience. We are unfamiliar to think in the direction of making good use of inconveniences.

Secondly, even if it were possible to make something inconvenient, it would be difficult to come up with a way to yield benefit from it. There is an inconvenient inn in Kyoto, Japan surrounded by rivers and mountains and inaccessible by land. The old map I used when I was a student shows "Arashiyama hot spring resort," but I did not even notice it existed when I was a student. Today, however, the inn at the same location is highly popular. There must have been a way to make the "inconvenient to access" into a "desire to visit," but it is not easy to come up with such a strategy.

Thirdly, it is difficult to evaluate design proposals. Even if a beneficial experience is envisioned, there are no criteria that can be relied upon in deciding whether to move to the implementation phase of the proposed design. When the idea of introducing the inconvenience of the same time bindings as the physical world into a healing game came up, I imagine that those who decided, "Okay, let's go with that," were not convinced that it would be the best-selling series.

Thus, in order to design BoI systems (to ideate systems that provide BoI to the user), it seems that I have to groan with a frown and folded arms in front of a difficult subject. On the other hand, under the slogan of "design thinking" (IDEO,

2022), a method of group work design by teams consisting of members with diverse backgrounds has become known. This method relieves us from the task of groaning while folding our arms alone.

When I conducted practical trainings and workshops on the theme of "designing BoI systems" following design thinking, we observed several characteristic thought processes. This chapter introduces them as methods of BoI system design.

2.2 Methods of Ideating Systems with Benefits of Inconvenience

As discussed throughout this book, when considering BoI systems, the two conceptual axes and four quadrants shown in Fig. 2.1 facilitate understanding.

In general, it is often thought that "if it is convenient, it is good enough." Let's express this idea by aligning the "convenience" axis and the "benefit" axis (if it can be assumed) pointing in the same direction or by being identical. Then, "convenience and benefit are independent," which is the premise when considering BoI, can be expressed by making these two axes orthogonal to each other.

In the planes defined by these two axes, the upper right and lower left quadrants, which are labeled "benefit of convenience" and "harm of inconvenience" in Fig. 2.1, are the place to plot well-known such cases as "convenience was good" and "inconvenience was unfortunate." On the other hand, the lower right quadrant, labeled "harm of convenience," is the place to plot cases in which convenience brings harm, while the upper left quadrant, labeled "benefit of inconvenience," is the place to plot cases in which benefits are derived from inconvenience.

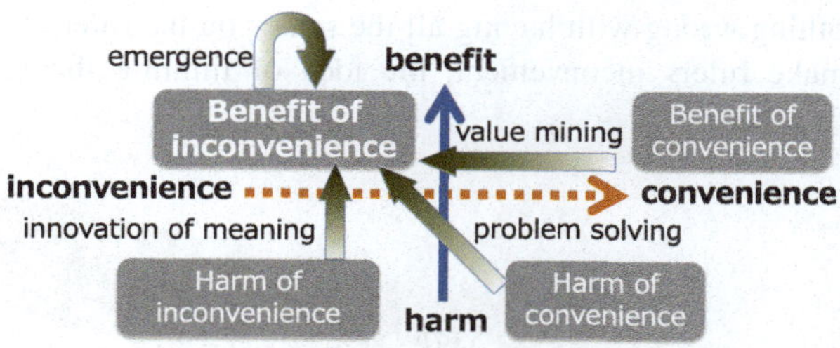

Fig. 2.1 The orthogonal relationship between convenience and benefit and four types of ideation methods. Unlike the idea that convenience is always beneficial, "convenience/inconvenience" and "benefit/harm" are considered to be independent and are represented schematically by orthogonal axes. Four quadrants are assumed in the plane defined by these two axes, which are abbreviated as benefit of inconvenience, harm of convenience, harm of inconvenience, and benefit of inconvenience, respectively. However, interval or ratio scales cannot be assumed for convenience/inconvenience. Two axes, their intersections, and the positional relationships of the planes shown in the figure are merely schematic representations of conceptual relationships

When ideating a new BoI system, which is a product or a service that provides users with the experience of gaining benefits because of inconvenience, the ideation process can be classified into four main types. They can be mapped to the four transitions of the four quadrants as shown in Fig. 2.1. This chapter names the value mining type from benefit of convenience to BoI (upper right to upper left), the problem-solving type from harm of convenience to BoI (lower right to upper left), the innovation of meaning type from harm of inconvenience to BoI (lower left to upper left), and the emergence type from BoI to BoI (upper left without transition to upper left).

2.2.1 Value Mining Type and BoI Card

By following two steps:

- Forcibly redesign a product, which is originally convenient with no problem, toward being inconvenient.
- Find (mine) new value in them.

We can ideate new BoI systems. This is called the value mining type thinking process. In the quadrants shown in Fig. 2.1, this can be viewed as a transition from a benefit of convenience (upper right) to BoI (upper left). When design workshops were held on the theme of BoI system design, it is one of the ways that the participants seem to enjoy coming up with ideas and sharing them. Making inconvenience seems to evoke a sense of mischievousness in the participants.

Figure 2.2 shows an example of a design ideated in the value mining type, which was born from the group work of several students at Kyoto University Summer Design School.

There is nothing wrong with having all the scales on the ruler, it is convenient. In order to make rulers inconvenient, the idea of limiting the scale to prime

Fig. 2.2 Example of idea by the process of value mining type: Prime Number Ruler. Scales are only available for prime numbers. Simple subtraction is required to accurately draw a line of length for unscaled numbers. It is sold as an original good of Kyoto University, Japan

numbers was proposed. From this inconvenience, such benefits as "enjoyment of the ingenuity of the measurement method" were mined.

However, it is generally difficult to come up with ways to make something inconvenient. Through the ideation process, our thoughts tend always directed toward making it convenient. In such cases, the BoI card, described in Sect. 2.3.1, is helpful.

2.2.2 Problem-Solving Type and Engineering Thinking

In engineering, we are taught that first and foremost, it is essential to correctly identify the problem to be solved. Even in design studies, according to IDEO, which popularized design thinking as a proper noun, designing is a kind of problem solving (IDEO, n.d.). This style of engineering thinking and design thinking can also be applied to BoI system design. We call this a problem-solving type of thinking process. In the quadrants shown in Fig. 2.1, this can be viewed as a transition from harm of convenience (lower right) to BoI (upper left). That is,

- If we find harm of convenience (a problem).
- Then adopt "making it inconvenient" as a means to solve this problem.

For example, the "barrier-aree" design (Fujihara, 2010) is widely known in Japan that originated at a daycare center. "Aree" is Japanese for "existence," so barrier-aree is a synonym for barrier-free. From the viewpoint of BoI design, it can be regarded as a result of the problem-solving type of design. That is, the convenience of barrier-free access posits the harm of depriving residents of the opportunity for a little daily unconscious physical training. Defining this situation as the target problem to be solved, barrier-aree design can be regarded as a solution for this problem by installing minor barriers in the facility of the daycare center on purpose.

2.2.3 Emergence Type and Hundred Knocks

The above two types are somewhat systematic since both follow a fixed process and have a common interpretation as transitions over the quadrant in Fig. 2.1. A style of thinking that differs from both is what we call emergence type of thinking. The process is:

- Cramming your heads full of BoI case studies, regardless of the field.
- Putting their essence under your consciousness without verbalizing them.
- Mass-producing ideas as you come to mind with folding your arms.

This type of thinking can be regarded as a type that returns from the upper left to the upper left without crossing over the quadrant as shown in Fig. 2.1.

This type of thinking requires us considerable skill. Not only the amount of knowledge and examples accumulated, but also a high ability to flexibly change them by analogy is necessary. The method can be seen as an application of what professional designers call "one hundred knocks" to BoI design. It is difficult but fun method that, if you can master it, will allow you to produce mass ideas that you never thought you would have.

2.2.4 Innovation of Meaning Type

The last remaining of the four quadrants shown in Fig. 2.1 is the harm of inconvenience (lower left), and we can also consider the type of thinking that starts here and transit to the BoI (upper left). This type of thinking turns "inconvenient and harmful cases" into "cases that are beneficial because they are inconvenient." One of the promising methods of this type is to "change the meaning" (Verganti, 2009).

As described in Sect. 2.1, an inn still stands where it was marked on the map as "Arashiyama hot spring resort" when I was a student. The inn transformed the "harm of inconvenience" of being surrounded by mountains and rivers that makes it difficult to access into a "benefit of inconvenience" as if it were an unexplored region. The inn now gains popularity.

It is interesting to know how they changed the meaning of inconvenience of inaccessibility from harm into a benefit. There must be special techniques, but nobody knows them except for the person who reconstructs the inn. Anyway, the fact is that the meaning of inconvenience was changed, so we call this type of design "innovation of meaning" and add one of the methods of BoI design.

2.3 Tools for Ideation of BoI Systems

2.3.1 BoI Card: How to Make Things Inconvenient and which Kind of Benefits Are Derived

In some cases, the BoI card shown in Fig. 2.3 can be effectively utilized when ideating BoI systems.

The contents shown on cards are based on the classification of examples of BoI systems from the viewpoint of "which inconvenience are observed" and "which benefit are derived" (Naito et al., 2015). The ways of making something inconvenient were classified into twelve and benefits derived from inconvenience were classified into eight. These classifications are displayed on cards and utilized as tools for supporting ideation (Hasebe et al., 2015).

As mentioned in the previous section, we are not used to thinking for make something inconvenient. In the case where we need to ideate inconvenient things

Fig. 2.3 BoI card. Yellow cards (light colored in B&W printings) show 12 ways of making things inconvenient that are derived from the classification of case studies of BoI systems. They are explained by pictograms on the cards and help us to ideate BoI systems. Green cards (darker colored in B&W printings) show eight kinds of benefits that are derived from some inconveniences. They also are results of the classification of case studies of BoI systems, are explained by pictograms on cards, and support our ideation

and matters, twelve types of "how to make it inconvenient" on the cards with yellow background (light color in B&W printing) in Fig. 2.3 will help you to ideate some inconvenient things that have potential to give users some benefits.

Next, when finding (mining) benefits from a design proposal that has been changed to be inconvenient, the eight types of "benefits from inconvenience" shown on the cards with green background (darker color in B&W printing) in Fig. 2.3 can be used as support.

2.3.2 *Necessary Conditions of BoI System*

As anyone can intuit, inconvenience does not always bring benefit. This is why the designer's creativity is required for a design that will be beneficial. In other words, inconvenience to the user does not mean that benefits will arise spontaneously. The designer must build into the design the mechanism by which the benefit occurs. The following checklist is provided to support such thinking:

- Both benefit and inconvenience are users.
- Benefits can only be gained by inconvenience.
- Benefits are not the primary objective.

These are only empirical rules, but they are what we consider necessary for what we call the BoI system. Answering "yes" to all three questions is a necessary condition for BoI design and is also an item to be checked when approving and rejecting design proposals. However, these are not sufficient conditions. Their satisfaction do not necessarily mean that the design can be called a BoI design.

Adding a little explanation, the first condition means that "your inconvenience is my benefit" is not appropriate to call it a BoI system, because it is the worst design, "I gain while others lose."

The second condition is not a negative situation in which there is only a confounding between inconvenience and benefit, but a positive situation in which there is a causal relationship between inconvenience and benefit, i.e., "there are benefits to be gained because of the inconvenience."

The third condition eliminates such statement as "abdominal exercises are hard and inconvenient, but they are beneficial for the abdominal muscles" from BoI. In contrast, such a statement as "abdominal exercise is more inconvenient than taking slimming pills, but it makes bodybuilding your own business," is the case of BoI.

2.4 Conclusions

This chapter describes the conception of an idea that introduces an inconvenience on purpose with the intention of obtaining the benefits of inconvenience. On the other hand, however, there may be a case where an inconvenience is adopted out of necessity, resulting in a BoI system. The cell production system is an example that brings BoI. The reasons why many manufactures adopted the cell production system at the end of the last century may such negative ones as "it was unavoidable in order to flexibly respond to high-mix low-volume production" or "the gain cannot be expected to be commensurate with the investment in line facilities" (Hashizume, 2006). However, we are told by workers there that the resulting gains are synergistic effects on worker motivation and skill development. BoI focuses on such positive aspects of human-included systems. This may not apply to those who decided to adopt the cell production system a quarter of a century ago, but when designing inconvenient products or matters in the future, I hope that these positive benefits will be (intentionally) foreseen and focused on, and innovation will occur in a way that makes benefits apparent.

References

Fujihara, S. (2010). *No need to be strong, just only live without being weak (ISBN 978–4–492-04386-8)*. Toyo Keizai. in Japanese.

Hasebe, Y., et al. (2015). Card-type tool to support divergent thinking for embodying benefits of inconvenience. *Web Intelligence, 13*(2), 937–102. https://doi.org/10.3233/WEB-150312

Hashizume, Y. (2006). Paradox of the cell production system. *Japan Society of Business Administration, 17*, 61–74. in Japanese.

IDEO (2022) Design thinking, https://designthinking.ideo.com/, last accessed 2022/09/27.

Japan Industrial Design Association (2022). https://www.jida.or.jp/, last accessed 2022/11/22.

Naito, K., et al. (2015). Design support method for implementing benefits of inconvenience inspired by TRIZ. *Procedia Engineering, 131*, 327–332. https://doi.org/10.1016/j.proeng.2015.12.396

Verganti, R. (2009). *Design driven innovation (ISBN 978–1422124826)*. Harvard Business Review Press.

Hiroshi Kawakami is a professor at the Kyoto University of Advanced Science, in Japan. He received his Bachelor, Master, and Dr. Eng. degrees from Kyoto University. He started his career at the Okayama University as an assistant professor. He joined Kyoto University, where he was an associate professor of the Graduate School of Informatics and a program-specific professor of the Design Unit. His research interests include systems design, where he has proposed FUBEN-EKI that stands for designing systems based on appreciating "benefits of inconvenience." He received the best paper awards from *Transactions of the Society of Instrument and Control Engineering* (1990, 2001, 2013), the *Transactions of Human Interface Society* (2009, 2018), and the *Journal of Society of Automotive Engineers of Japan* (2014).

Chapter 3
"Fuben-eki" in Service Design

Yasuo Nosaka

Abstract Based on my research on corporate marketing measures over the years, I continued to collaborate with Ph.D. Hiroshi Kawakami on the application of the technology to the inconveniences service design field. As a result, seven inconveniences, such as "increase of number of operation" and "less information," and seven benefits, such as "feel affirmative" and "be fun," were extracted. We also introduced the concept of "Nudge" in the sense of "gently pushing them back" what kind of "Fuben-eki" arose in the processes of influencing factors, marketing measures, and products/services. By showing this in the structure diagram, the company succeeded in creating 20 types, including "value co-creation." Based on the analysis of the "Fuben-eki" gained in this way, we explored the utility of the inconveniences in service design by drawing on the <inconveniences→benefits> seen in various service design areas, including product service development, space business, SDGs, human resources strategy, Gen Z, and democracy. In the future, we will apply this knowledge to the design of "Fuben-eki" measures and to the introduction of VoC data analysis.

Keywords Benefit of inconvenience · Service design · Nudge

3.1 Introduction

"Fuben-eki" is a thought of excavating the value positively from inconvenience. According to Ph.D. Hiroshi Kawakami, who advocates the concept of "Fuben-eki," this thought was originally proposed in the field of system design. Subsequently, R&D began in the field of product design, starting with "Prime number ruler."

Considering that this could be applied to the field of service design, Hakuhodo and Ph.D. Kawakami began joint research. I would like to summarize the research

Y. Nosaka (✉)
Hakuhodo Co., Ltd., Akasaka, Minato-ku, Tokyo, Japan
e-mail: yasuo.nosaka@hakuhodo.co.jp

H. Kawakami (ed.), *Systems Design Based on the Benefits of Inconvenience*,
Translational Systems Sciences 31, https://doi.org/10.1007/978-981-19-9588-0_3

results that have been carried out for more than 2 years, the applications to each field, and the outlook for the future.

3.2 Approach from Corporate Marketing Measures Research

As a worker at Hakuhodo, I learned about Kawakami's advocated "Fuben-eki" when he introduced a book in 2018, entitled "The idea of Fuben-eki," (Kawakami, 2017) which he wrote.

At that time, I was in the Department of the Research and Development Bureau and was in charge of collecting and analyzing useful knowledge related to the advertising business. One of these was the case studies of corporate marketing measures, which had been conducted for many years, and my supervisor suggested that it would be possible to take advantage of the "Fuben-eki" to corporate marketing measures.

Originally, however, corporate measures are undertaken to provide "convenience benefits" such as usability and rationality to customers (or users). As I read his books (Kawakami, 2017; Kawakami et al., 2017), I wanted to make a plan measures based on this idea. Currently, the pursuit of "convenience benefits" is progressing further, as exemplified by the fact that most of all corporate marketing measures are completed with smartphone apps.

However, the more efficiencies and rationalizations are pursued regardless of industry, the more companies are faced with the problem of not being engaged in by consumers (i.e., being unable to attract consumers' attention, and being unable to be regarded as themselves). So, I thought that the customer approach of "Fuben-eki," which indicates the benefit of inconvenience, would also be effective.

That is why, having felt as if searching for sand from a sandy beach, we reviewed the measures taken by each company in various industries over the past 20 years to see if there had been any measures that could find such "Fuben-eki" in the past. And we found a few dozen of such measures that had involved it somehow. Next, we attempted to classify these inconvenient and benefit factors according to the "inconvenient card" and the "benefit card" described in Ph.D. Kawakami's book. In some cases, there were notable benefits such as "fun" and "grad" that were peculiar to the corporate marketing measures. These were also added as items of benefit, and we made a prototype matrix to deal with inconvenience and benefits and sent an e-mail. This led to the start of a relationship with Professor Kawakami, who was a teacher at Kyoto University at that time, and in April 2019, a joint research was officially started in the form of an industry–university collaboration between us (Hakuhodo) and Kyoto University.

Since then, for more than 2 years, we have been engaged in research to identify principles from various past corporate marketing measures cases, discuss them, and conduct structural analyses. First, we reviewed the "inconveniences" and "benefits" that constitute "Fuben-eki." In order to design a marketing strategy that incorporates Fuben-eki, the most important aspect is what kind of means (inconvenience) and how to motivate customers and users (benefit). To do so, as Ph.D. Kawakami noted in his book, it was necessary to scrutinize the objective–means relationship regarding inconvenience and to distinguish between objective and subjective (emotional) aspects regarding benefits. In the analysis, a sample of 20 Fuben-eki corporate marketing measures was analyzed into seven instrumental inconveniences and seven emotional benefits shown in Fig. 3.1 (Nosaka et al., 2019).

3.3 The Structural Analysis of the Fuben-eki in Service Design

Next, we clarified the structure of corporate marketing measures (Fuben-eki measures) that can provide Fuben-eki to customers, and clarified which of these processes results in the Fuben-eki (Nosaka et al., 2020; Kawakami et al., 2020).

Corporate marketing measures are basically implemented to promote the continued purchase and use of products and services and to increase loyalty by eliciting and enhancing the attractiveness of the company's products and services. Naturally, there are inconveniences and benefits inherent in the products and services themselves, and there are also inconveniences arising from corporate marketing measures and their relationship. In doing so, we must also consider Influencing Factors such as the "COVID-19 crisis." This is because this factor may cause inconvenience. In other words, when considering Fuben-eki in "corporate marketing measures," it is necessary to consider factors such as "Products and Services" and "Influencing Factors."

In addition, the concept of "Nudge" has been introduced to form a framework for expressing the structure of "Fuben-eki" in corporate marketing measures in a uniform manner.

The term "Nudge" has recently become widely known as a term used in behavioral economics, but it has long been known in the field of design as one of the tricks and tips and is described as a "method of promoting behavior without limiting options or providing special rewards." (Lidwell et al., 2015).

In "Fuben-eki measures," customers generally do not necessarily feel the benefit just because they feel inconvenient. On the contrary, some customers unconditionally avoid situations that are inconvenient. This is why drivers are included in the Fuben-eki measures to encourage customers to move from inconvenience to benefit, which is what makes it appropriate to call "Nudge."

This "Nudge" has the following five principles (conditions) (Lidwell et al., 2015):

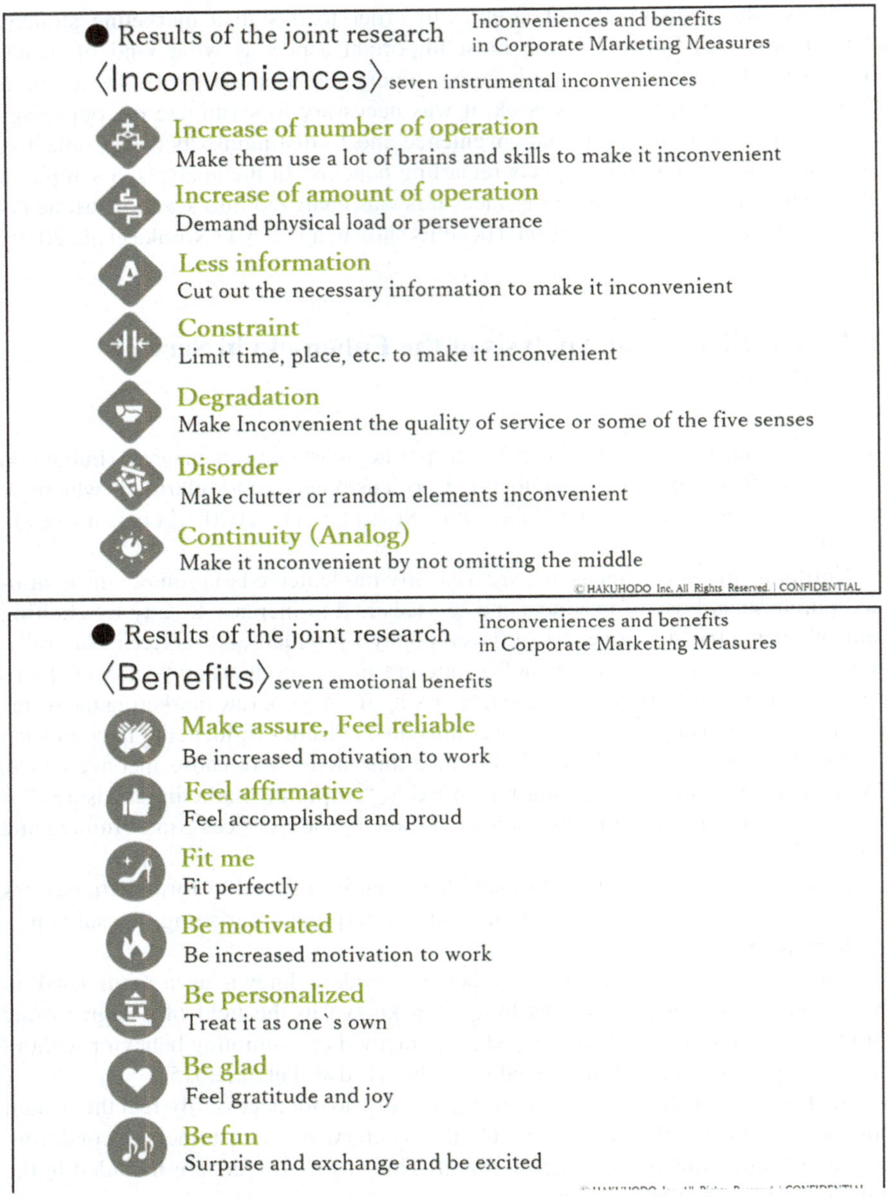

Fig. 3.1 Seven instrumental inconveniences and seven emotional benefits

1. **Convenient default (default)**
   The first setting is to choose the most convenient and most desirable option, rather than the least harmful one.
2. **Clear feedback**
   Provide visible and clear feedback on actions immediately.
3. **Remuneration (Nature of mankind) directly related to objectives**
   Rewards lead directly to desirable actions and do not create situations that are in conflict with each other.
4. **Structuring of alternatives**
   To facilitate decision-making by providing a means to clean up and simplify complex issues.
5. **Visualization of goals**
   Create a mechanism for easily measuring performance and display it at a glance. Then people can then immediately assess their achievements against the goals.

The structure of the Fuben-eki can be illustrated by considering how the "Nudge" that satisfies these conditions is driving the "inconvenience" of the Fuben-eki measures.

For example, if a member takes a picture of the sky above him/her and sends it to his/her club, it gathers from all over his/her country, so the club gets an accurate picture of the weather, and the result is returned to the member. It takes much effort for members to take pictures of the sky at their locations and send them to the club. That is why detailed information on the whole country is collected, making it possible to obtain a more accurate picture of the weather (Nosaka et al., 2019). Not only is this in the public interest but the weather in the vicinity with members can be predicted more accurately. It is hard to get people to take pictures and send them to us, but if they want the Web community to be happy about the sky, they will be able to accept it in their favor. We classified the relationship between inconvenience and benefit as "value co-creation." "Value-co-creation type" is based on the previous analysis of the type of Fuben-eki measures in which one cooperates with a company's services so that one can enjoy the benefits better.

The present example is shown in Fig. 3.2. Seven instrumental inconveniences shown in Fig. 3.1 for influencing factors, corporate marketing measures, and products/services are shown in yellow slots (three middle-row boxes) and seven emotional benefits are shown in green slots (three bottom-row boxes), and the relationship is explained by the up-down arrows. The inconvenience of corporate marketing measures depends on the "Nudge" that supports it. The Fuben-eki of corporate marketing measures are classified into around 20 types, but one of them is "value-co-creation type."

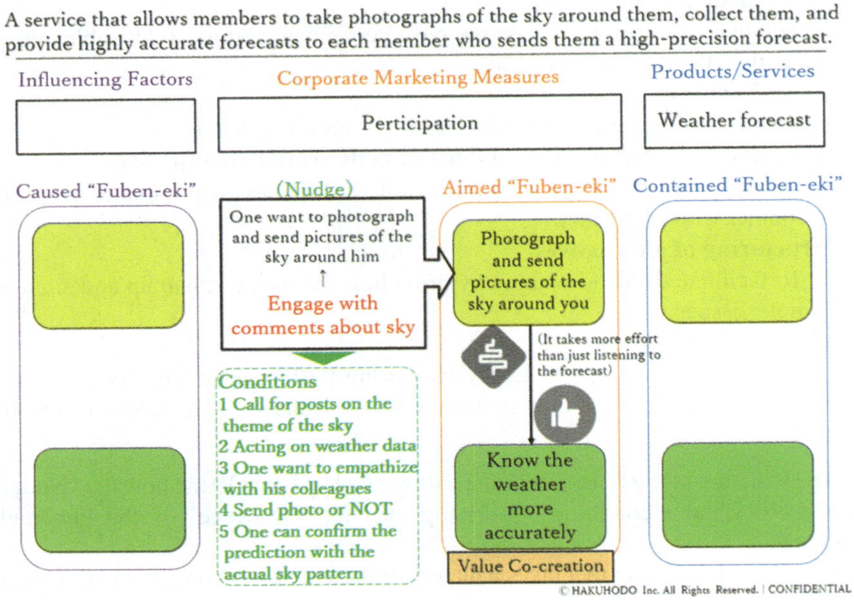

Fig. 3.2 The example of the type (Value Co-creation) of Fuben-eki measures

3.4 Application of Fuben-eki Analysis to Various Service Designs

We have thus proceeded with the analysis of Fuben-eki based on corporate measures, but these can be considered in the broader area of service design. The Fuben-eki can be applied in a variety of contexts described below. This chapter discusses it in various categories.

3.4.1 Product Service Development and Fuben-eki

The Fuben-eki includes not only corporate marketing measures but also the development of UI/UX embodied in products and services. This is another service design.

In fact, I have worked with a home appliance manufacturer, but many of the appliances that I have used so far have become more convenient, able to do everything automatically, and have added so much functionality that people cannot keep up with them that they have promoted commoditization or become unable to fully use them. As a result, price competition intensified, and the market was lost to single-function products. This meant that executives with a sense of crisis were trying to find a way out of their inconvenience.

Although I will not go into detail, it was shared that the design of Fuben-eki, which elicits objective benefits such as creativity, discovery, and understanding of the target system was useful for the purpose of realizing the important value of fostering rich of human relationships through the use of home appliances and the rest of the fun of using home appliances.

3.4.2 Space Business and Fuben-eki

Space is a treasure trove of inconvenience. We are forced to live a tight life in outer space. The data obtained from satellites is from a panoramic perspective of the earth, but there are various limitations (SORABATAKE, 2021).

To convey the appeal of the universe, we often see it creating a theme park-like pleasure, but the universe is a place with many elements, such as adventure and exploration. This is where we want to actively apply Fuben-eki. In fact, the "constraint that cannot be used smoothly" brings many possibilities to people (Kawakami, 2021). Using the seven instrumental inconveniences and the seven emotional benefits shown in Fig. 3.1, the relationship between inconvenience and benefit in Fuben-eki is shown below.

$$\text{Deterioration} -> \text{Be motivated}$$

Focusing on "inconvenience" without the atmosphere (oxygen), water, and gravity, which are naturally present on the earth where there is enthusiasm to do so, we can reconsider outer space as a living device that can obtain ideas for life. By doing so, we may be able to think about new possibilities, such as lifestyle changes, from the outset, and we may be able to use them in various corporate measures.

Indeed, even within us, there are plans for demonstration experiments on whether we can apply Fuben-eki theory to outer space data and promote behavioral change in humans.

3.4.3 SDGs and Fuben-eki

The SDGs (Sustainable Development Goals) are sustainable development goals whose essence is often described as "three good" by Omi merchants, known as representative examples of management philosophy in Japan (Oki et al., 2018). In other words, this concept satisfies the environment, society, and economy and extends it spatially and temporally (Oki et al., 2018). This is what I call "pan-marketing," because there is an expanded exchange of satisfaction, a market is formed, and a business is established.

Part of this can only be realized with inconvenience. It is a kind of "value-co-creating" corporate marketing measure that can only be achieved with the cooperation of customers, such as the introduction in Chap. 3, in which weather forecasts can be obtained with greater precision by having customers send a photo of sky. Not only do the benefits of this become the public interest, but it also gives members a more accurate indication of the surrounding weather, which makes them feel that their actions have been useful and that they have been able to gain a sense of self-affirmation that they have done secondary good things. In other words, they satisfy the environment (precise understanding of weather), society (public interest), and the economy (corporate and member benefits). In this way, value-co-creating corporate marketing measures are also effective in the context of the SDGs. The relationship between inconvenience and benefit in Fuben-eki shown in Sect. 3.4.2 is as follows.

Increase the number of operations– > Feel affirmative

Let me introduce another. A detergent manufacturer has implemented a policy of collecting and donating clean clothes to be sent to developing countries. At this time, the participants have to bring their donated clothes to the venue.

But they can make sure that the profitable clothes they donated become clean and neat by the manufacturer's detergents. In other words, they are proud of their contributions to society without feeling guilty. Of course, the manufacturer's awe and gratitude for the product will also be felt. This is a so-called "witnessed and realistic" corporate marketing measures, and the SDGs have been established. The relationship between inconvenience and benefit in Fuben-eki is as follows:

Increase the amount of operation– > Feel affirmative

3.4.4 Human Resources Strategy and Fuben-eki

The four quadrants shown in Fig. 3.3 are often used in conceiving Fuben-eki (Kawakami et al., 2017). In other words, they are classified into four groups based on the following categories: "convenience" or "inconvenience" and "benefit" or "harm."

- **"Convenience Benefits" - > "Inconvenience Benefits" i.e., Fuben-eki**
  (Uncover new benefits by making them inconvenient)
- **"Convenience Harms" - > "Inconvenience Benefits" i.e., Fuben-eki**
  (Solves problems arising from convenience by making them inconvenient)

It is about thinking about the Fuben-eki in such ways.

I would like to add the time horizon to these. We can see that there are things that become inconvenient in the changing times, or things that are harmful.

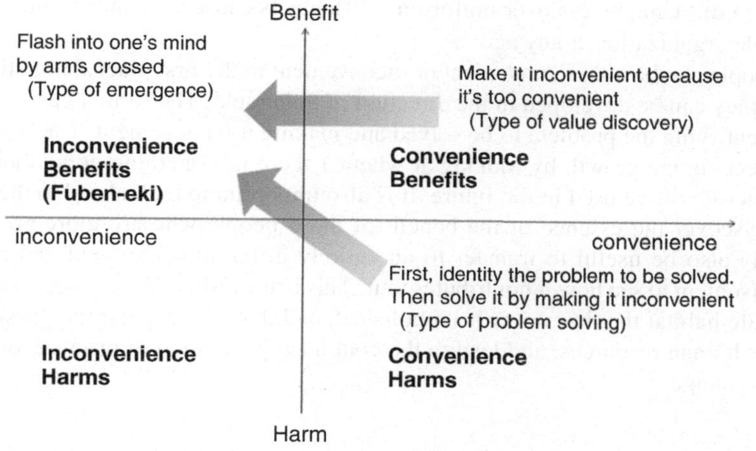

Fig. 3.3 The four quadrants often used in conceiving Fuben-eki (Kawakami et al., 2017)

For example, abacuses, which used to be a convenient tool for calculation, have been regarded as a time consuming and inconvenient tool since the advent and spread of calculators over time.

When plastic straws appeared at a time when only thatching straws and paper were available, they were really beneficial. However, when the environmental degradation caused by plastics was considered a problem over time, they were regarded as targets to be reduced. This is why it is important to keep the idea of "Fuben-eki" in mind.

The "Convenience Benefits" of abacus should be thought of as a Fuben-eki, not" Convenience Harms," by seeing that it will eventually change to inconvenience, and finding its inherent utility (the ability to cultivate computational power and internalize its ability in the form of mental arithmetic). The relationship between inconvenience and benefit in Fuben-eki is as follows:

Increasing the number of operation− > Feel affirmative

Plastic straws can be solved by remembering tools that existed before they were targeted for reduction and restoring lost abundance (straw and paper eventually returned naturally) to a modern age with advanced science and technology. The relationship between inconvenience and benefit in Fuben-eki is as follows:

Degradation− > Make assure, Feel reliable

I think that HR strategy is the best way to take advantage of this concept. In the case of personnel education, it may not be the best simply to develop everyone into convenient and profitable human resources. This is because, as time goes by, it may turn into an inverse evaluation, as in the case of abacus or plastic straws. We have a motto of "Uniqueness over uniformity." This is because we can continue to be a flexible organization at any age.

People can be either convenient or inconvenient in the first place, but at the very least they can be developed in the direction of Fuben-eki. The former can be solved by identifying the problem to be solved and making it inconvenient. The latter may well encourage growth by looking at whether there is something good about it or whether it will be used in the future. It is also important to look ahead to the future and discover the essence of the benefit of those people who are more profitable. It may also be useful to transfer to an entirely different department and make it inconvenient to see how it contributes to it. This is the utility that has been hidden in the safe habitat that has never been polished, and that will expand the possibilities for its human resources, and lead to the continued generation of profits even as the times change.

3.4.5 Generation Z and Fuben-eki

The Fuben-eki is an antithesis of convenience and rationality, which does not seem to fit into the young people represented by generation Z (Gen Z).

However, there are some trends that are not in fact. For example, some apps can simulate analog cameras around the 1960s by manipulating smartphones like film cameras. However, not only can analog photos be taken, but the fact that they realistically reproduce the "feeling of using film cameras" is becoming popular among young people. There was a revival boom for film with lens a few years ago, but this generation seems to be more attracted to analog items on the contrary because it can do anything with a single smartphone. The relationship between inconvenience and benefit in Fuben-eki is as follows:

$$\text{Degradation} -> \text{Be personalized}$$

In addition, there is a social game for smartphones, in which cards featuring popular characters appear during play are popular. Instead of simply buying a ready-made card, they are making their own portable goods, and looking at them at the hard times at part-time job and other places. It is meaningful to make this card yourself. However, the inconvenience favored by some people may vary because they want to cut and create an existing picture (the kit should be assembled) or they want to make an original picture (in this case paper and brush should be provided). The relationship between inconvenience and benefit, in this case, is as follows:

$$\text{Increase the number of operations} -> \text{Be personalized}$$

Ph.D. Kawakami, who was mentioned earlier, also reported in a magazine that women like inconvenience more than men, and the younger more than the elder (Kawakami, 2021).

3.4.6 Democracy and Fuben-eki

Democracy takes much more time and effort than dictator decision-making. However, this is precisely why the checking function works, so the results chosen are more reliable and reliable.

Elections are at the heart of the process, but they are always cumbersome and require a vote at the polling station. Electronic voting seems to be less time-consuming, but in this case it is easy to vote according to mood. The current election system is indeed Fuben-eki.

It is an cumbersome process that reaffirms its importance each time. Instead of leaving it to others, we can say that it gives rise to the power of self-determination (Yamaguchi, 2021). It takes a lot of time and effort, but the results are more reliable because the control function also works. Above all, we can decide our own destiny-democracy is indeed "Fuben-eki." The relationship between inconvenience and benefit in Fuben-eki is as follows:

Increasing the number of operations– > Make assure, Feel reliable

3.5 Further Outlook and Challenges

3.5.1 Design of Fuben-eki Measures

Chapter 3 introduces the structural analysis of Fuben-eki measures. This can be used to conceptualize and design new Fuben-eki measures. The idea is that it may not be difficult to develop methods using conventional materials. In this case, however, the inconvenience should also be considered. This can lead to reactions that are different from what is intended.

This design cannot be made overnight. Deep insights and analytical capabilities will be required. Chapter 4 looks at the Fuben-eki in a variety of contexts, but this is actually the case.

We believe that the first step will be to conduct experiments and observations to capture the Fuben-eki of service design in a wide range of areas and to optimize them.

3.5.2 Introduction to VOC Data Analysis

VOC (Voice of Customer) can also be viewed as a trove of Fuben-eki, not just a claim. Usually, such a voice is an attempt to solve the inconvenience with convenience benefit, but it takes considerable energy to do so. There are even things that do not work out. Therefore, we will also consider the solution by Fuben-eki.

If we can accumulate solutions by "Fuben-eki," not only can we know how to do it, but we can also do machine learning so that we can respond appropriately. The benefits of large-scale VOC data analysis by AI will be incalculable. With this in mind, I would like to continue R&D in the future.

Acknowledgments The drafts were greatly assisted by Professor Hiroshi Kawakami, as well as Koji Kato and Kota Ito of our Joint Research Team on Fuben-eki, Industry, and Academia. Mr. Yusaku Konno, the group manager who introduced me to the Fuben-eki, would like to take this opportunity to thank me.

References

Kawakami, H. (2017). *The concept of "Fuben-eki"*. Impress Corporation.

Kawakami, H. (2021). *The joy of things that "inconvenience" awakens, WORK MILL with Forbes Japan ISSUE 06*. OKAMURA CORPORATION.

Kawakami, H., Hiraoka, T., Kokita, M., et al. (2017). *Fuben-eki*. Kindai kagaku sha.

Kawakami, H., Nosaka, Y., Kokita, M. (2020). Structural analysis of measures that benefit from inconvenience. In: The Society of Instrument and Control Engineers, Systems and Information Division, SS5–1-3.

Lidwell, W., et al. (2015). *The pocket universal principles of design: 150 essential tools for architects, artists, designers, developers, engineers, inventors, and makers*. BNN.

Nosaka, Y., Kawakami, H., Kokita, M. (2020). Structure analysis of benefit of inconvenience in measures by corporations. In: The Institute of Electronics, Information and Communication Engineers, IEICE technical report, 120–190, MVE2020–26, pp. 25–30.

Nosaka, Y., Kawakami, H., Momose, M., et al. (2019). A thought of idea-method using "Fuben_eki" for planning measures by cooperations. In: The Institute of Electronics, Information and Communication Engineers, IEICE technical report, 119–190, MVE2019–16, pp. 61–66.

Oki, T., Onoda, S., Kuroda, K., et al. (2018). *Fundamentals of the SDGs*. Graduate School of Project Design Press.

SORABATAKE (2021) Homepage, https://sorabatake.jp/20336/, last accessed 2021/5/28.

Yamaguchi, S. (2021). *Technical Liberal arts for freedom*. Kodansha.

Yasuo Nosaka Graduated from Waseda University, Department of Electrical Engineering, Department of Science and Engineering, 1989.

He joined Hakuhodo Co., Ltd. in the same year.

After working at the Business Bureau and the Research and Development Bureau, he is currently in the Business Development Bureau.

Chapter 4
BoI Value: The Third Value in Value Engineering (VE)

Manabu Sawaguchi

Abstract VE (Value Engineering) is a design method for improving value through a good balance between function and cost. Conventionally, VE has two types of value: "Use value" in practicality, and "Attractive value" focusing on appearance and beauty. On the other hand, in Modern society (that is, "Social growth period"), where individual respects self-actualization, users themselves are willing to create "Inconvenience" by spending physical effort or psychological cognitive resources. But they sometimes think that this inconvenient activity is not so big burden, may determine that there is a "benefit" from the aspect of mental satisfaction. Positively considering such individual subjective evaluation, the author decided to define such satisfaction as a completely new "Third value" called "Benefit of Inconvenience (BoI) Value." Therefore, this paper systematically organizes and introduces the characteristics of the "Third value" to be built here, using the Kano Model, which is well known in the field of product development.

Keywords Value Engineering · Third value · Value in Benefit of Inconvenience · Kano Model

4.1 Features of Value Engineering (VE)

VE is one of the management techniques born in the United States in 1947, and it stands for Value Engineering. Prior to VE, IE (Industrial Engineering) originated from Taylor's "Scientific Management" (1911) and QC (Quality Control) presented in "The theory of Control Chart" (1924) by Shewhart have been developed in the United States (Yoneyama, 1969). The characteristics of IE and QC are to analyze problems in detail by focusing on the current structure and format of existing products and services. They go through a process of looking into countermeasures

---

M. Sawaguchi (✉)
Graduate School of Technology Management, Ritsumeikan University, Ibaraki, Osaka, Japan
e-mail: sawaguch@fc.ritsumei.ac.jp

H. Kawakami (ed.), *Systems Design Based on the Benefits of Inconvenience*,
Translational Systems Sciences 31, https://doi.org/10.1007/978-981-19-9588-0_4

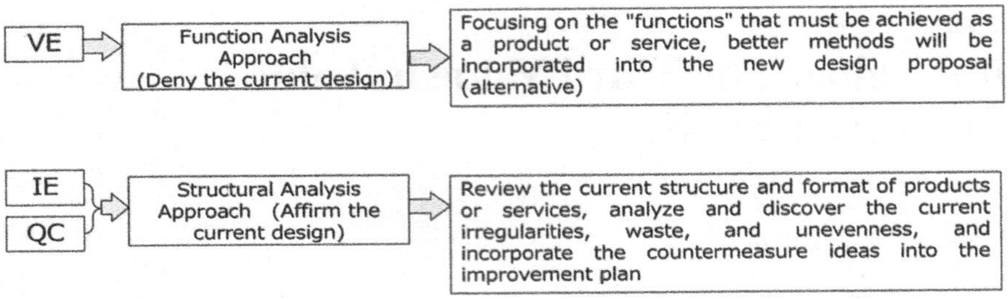

Fig. 4.1 Comparison of functional analysis and structural analysis approaches. Partial additions and corrections in reference (Sawaguchi, 2015)

and leading toward a proposal for a solution. Therefore, it is mainly useful in production management and quality control on the premise of mass production in the manufacturing industry. On the other hand, VE takes the approach of clarifying the "function" that a product or service should originally perform and creating a new design from that function. Therefore, VE is a particularly effective management technique for design-type problems, and the characteristics of these two approaches can be systematically summarized as shown in Fig. 4.1.

In other words, the greatest feature of VE is nothing other than "function-oriented thinking" and "expecting the creation of alternative ideas." In addition, the functions in VE refer to "the functions of products and services intended by design," and the intention is the work and benefit desired by the user.

4.2 The Role of VE up to the Present

IE/QC was introduced in earnest in post-war Japan, and greatly contributed to the realization of "powerful worksite = highly productive mass production processes" in Japanese manufacturing industry during the high economic growth period. On the other hand, many companies have introduced VE during their stable growth period after the oil shock (Tsuchiya, 1982). Specifically, VE has been recognized as an effective management technology for cost reduction, mainly in the export-based manufacturing industry (e.g., home appliances, automobiles, and precision machinery), which was struggling to reduce costs in an environment of strong yen and dollar depreciation, and VE aims to realize cost-effective design proposals based on the function-oriented thinking regardless of the existing structure. It was a timely method (means) for the corporate environment at that time.

4.2.1 *Functions Covered by the Current VE*

Since the 1980s, VE has not only been a means of cost reduction but also has been attempted to be applied at the upstream stages of product planning and development with the aim of realizing high added value. For this reason, although the target of VE is mainly "use function," there have been more and more cases where "Attractive function" related to design and aesthetics which make users want to own is addressed for products such as durable consumer goods aiming to improve added value. That is, the functions covered in the current VE are these two types.

When applying VE at the upstream stage, as not only products but also service areas and business models are addressed lately, use functions are subdivided into needs functions and want functions, and Attractive functions are subdivided into art design functions and letter functions, so that the functions necessary for users can be understood in detail (see Fig. 4.2) (Sawaguchi, 2020).

4.2.2 *Relation Between the Quality in the Kano Model and the Function in VE*

In the fields of QM (Quality Management) derived from QC and product development, "5 categories of quality attribute" (see Fig. 4.3) developed by Dr. Noriaki Kano (Professor Emeritus, Tokyo University of Science) in 1980 are widely known (Kamiesu, 2015; Japan Science and Technology Federation, 2021).

Particularly, "The Kano Model" (see Fig. 4.4) describes the changes in which each quality is achieved based on the priority of "must-be quality - > one-dimensional quality - > attractive quality (3 important quality elements)" on two

	Functions which customers desire (Sub-classification)	Example (Smartphone)
Use Function (Practical Function)	<Needs function> Function that makes a direct contribution, especially related to the core of the product, among other practical functions	"Enables voice communication" "Enables transmitting information through social media" "Take photos - Send and receive photos", etc.
	<Wants function> Function that contributes to further customer satisfaction with practical functions other than needs function	"Download your favorite APP" "Enjoy (selected) games", etc.
Attractive Function (Sensitive Function)	<Art design function> Attractive function in design aspect (color, shape, texture, etc.) that makes you want to own the product even more, including functions that appeal to the customer's vision, logo, etc.	"Thin and smart shape" "Bright and colorful case", etc.
	<Letter function> Attractive function in terms of naming and catchphrase that makes you want to own the product even more, and that mainly appeals to the customer's sense of words	iPhone 12 Pro, iPhone 12 Pro Max Galaxy S21 series 5G, etc.

Fig. 4.2 Classification and characteristics of functions targeted by VE. Partially modified (Sawaguchi, 2020)

Three important quality factors	**Must-be Quality** → Dissatisfied when being unsatisfactory; satisfied but taking it for granted	Call audio (It should be clear obviously, if difficult to hear customers will complaint)
	One dimensional Quality → Dissatisfied when being unsatisfactory; satisfied when fulfilled	Battery life (customers are satisfied if the operating time is long, but dissatisfied if it is short), weight, etc.
	Attractive Quality →There is no point in being unsatisfactory (I don't feel dissatisfied), but if it is satisfied, I will be happy	Hi-resolution sound sources (nice to have, but customers are OK without it), curved LCDs, etc.
Indifferent Quality →Whether satisfied or not has no impact on customer satisfaction level		No impact on people who are indifferent to voice input, such as Siri on smartphones
Reverse Quality → When satisfactory, customers become dissatisfied, and when unsatisfactory, customers are happy		Navigation with voice guidance at all times (drivers who want to listen to music carefully are dissatisfied with that function)

Fig. 4.3 Five categories of quality attribute advocated by Kano. Source: Reference Sawaguchi et al. (2020)

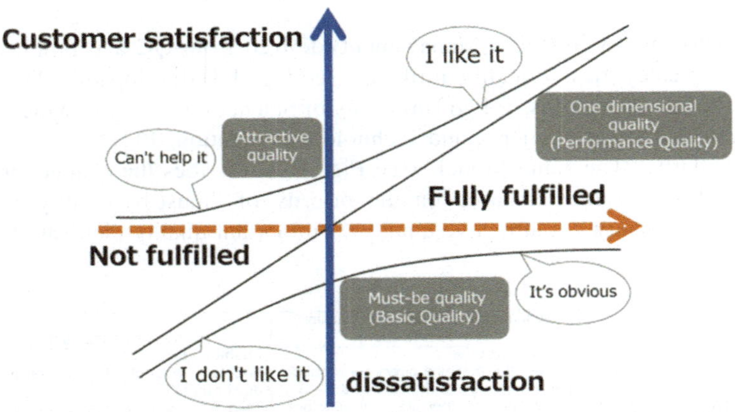

Fig. 4.4 The Kano Model, which shows customer satisfaction. Source: Reference Sawaguchi et al. (2020)

axes in response to high-end users is also internationally well known (Japan Science and Technology Federation, 2021).

According to the nature of the graph shown in this model, when the three quality elements that create value are added, the user's satisfaction keeps increasing steadily. That is, the three quality elements tend to increase monotony just like the two functions targeted in VE, thus they have an affinity for each other. Therefore, the author attempted to create a relative diagram based on the characteristics of the three quality elements and the two functions (see Fig. 4.5).

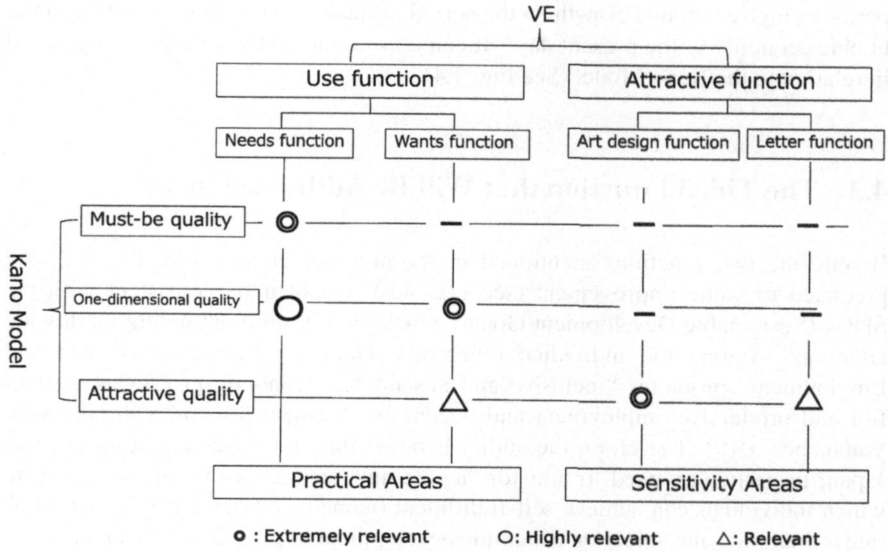

Fig. 4.5 Relationship between the three quality elements and the functions addressed in VE

Must-be quality and one-dimensional quality correspond to the use function in the practical domain. Moreover, it can be said that the must-be quality corresponds to the needs function, while the one-dimensional quality corresponds to the needs function when it is dysfunctional but then it corresponds to the want function after the function is fulfilled. On the other hand, the attractive quality is strongly related to the attractive function in the sensitivity domain, especially the art design function. It can be a want function (e.g., utility option for high-end users) or a letter function (e.g., original logo). Considering the importance of the design aspect seen in durable consumer goods, there is no sense of incongruity in general.

4.2.3 *Value-Enhancing Patterns at which the Current VE Is Aiming*

In the current VE, except for (1) value improvement by cost reduction (↑) V = F (→) / C (↓), all of them are value improvement by adding new functions and increasing the achievement level of existing functions, that is, three patterns that increase user satisfaction; (2) Realize cost reduction by increasing functions (↑) V = F(↑) / C(↓); (3) Maintain cost and improve functions (↑) V = F(↑) / C(→); (4) Cost is increased but functions are improved beyond that (↑) V = F(↑↑) / C(↑) are the main premise. This phenomenon (high-end pursuit model by adding new functions) is "the basic pattern of value improvement aimed at by VE from the period of high economic growth - > the period of stable growth - > the bursting of the bubble economy to the present day." It can be summarized as follows if organized in relation to the Kano Model (See Fig. 4.4).

4.3 The Third Function that Will Be Addressed in VE

If only the two functions mentioned in the previous chapter (see Fig. 4.2) are predicted to value improvement (see Fig. 4.6), would it be consistent with the SDGs (Sustainable Development Goals) which are currently attracting worldwide attention? Among the individual objectives, Goal 8, "Economic Growth and Employment" promotes "inclusive and sustainable economic growth as well as full and productive employment and decent employment for all." (Murakami & Watanabe, 2019). Therefore, the author believes that developed countries such as Japan, in particular, need to aim for "a safe and secure society and a society in which individuals can achieve self-fulfillment (named as "social growth period")" rather than focusing solely on economic development in pursuit of convenience.

4.3.1 Quality and VE Expected in the Social Growth Period

Considering the characteristics of social growth period, it can be said that "the society is shifting from tangible material consumption to intangible experiential consumption." In such an era, there is a limit on solely using monotonically increasing manufacturing approach through three quality elements and two functions as previously described. The author found interesting about the concept of "the benefit of inconvenience" proposed by Kawakami et al. (Kawakami, 2011) and focused on its potential of being as the third function in VE. The benefit of inconvenience can simply be described as "an act of intentionally devoting physical

Patterns of Value Improvement (Value-up)	Correspond to cost reduction	Almost corresponds to the "improvement of 3 quality elements (must-be quality, one dimensional quality, and attractive quality)" of Kano Model		
	①	②	③	④
	$\uparrow V = \frac{F \rightarrow}{C \downarrow}$	$\uparrow V = \frac{F \uparrow}{C \downarrow}$	$\uparrow V \frac{F \uparrow}{C \rightarrow}$	$\uparrow V \frac{F \uparrow\uparrow}{C \uparrow}$

Fig. 4.6 Three quality elements and value improvement patterns of VE. Source: Reference Sawaguchi et al. (2020)

burdens and psychological cognitive resources (inconvenience) but obtaining the user satisfaction more than that (convenience)." In other words, an attempt to intentionally make inconvenience and create benefits for users. "BoI (Benefit of Inconvenience) Function = Third function" in VE logically makes sense. (The methodology will be described later.)

On the other hand, the limit of conventional VE is a kind of contradiction in which the intention of intentionally realizing "benefit of convenience aimed at improving user satisfaction (benefit) by not spending physical labor or psychological cognitive resources (convenience)" versus "harmful convenience that leads to a decrease in user satisfaction (harm)" occurs.

In fact, because Kano had foreseen such contradictions from the 1980s, he must have also presented "indifferent quality and inverse quality (see Fig. 4.3)." If these five quality elements are integrated and organized in a two-axis graph of the Kano Model formula, they can be roughly illustrated as shown in Fig. 4.7.

Customer satisfaction by integrating each quality element is close to an S-curve, and at the turning point, a limit of indicating customer satisfaction by the growth of physical fulfillment on the horizontal axis can be seen. Of course, even from the viewpoint of conventional VE, we have tried not only physical satisfaction centered on use function but also the increase in attractive function in the sensitivity area through consumer consumption products. However, this second function does not intentionally interfere (creates inconvenience) with the use function, so it is completely different in nature from the BoI function which claims to be the third function.

Based on what has been discussed so far, the characteristics of the turning point (see the right graph in Fig. 4.7) can be summarized as follows:

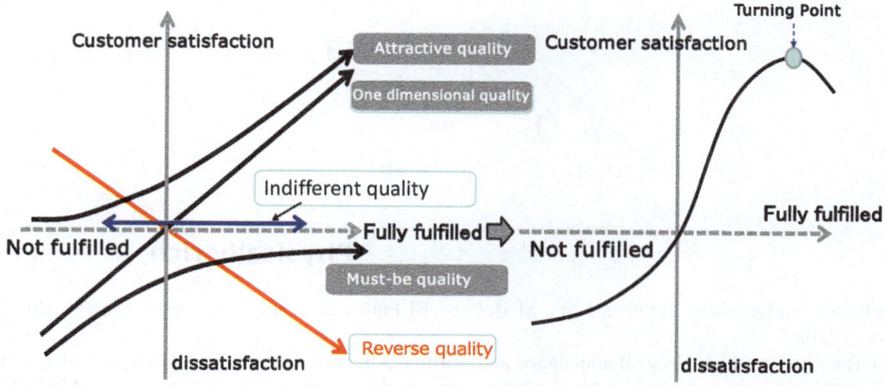

Fig. 4.7 Integrated customer satisfaction curve of Kano model and all quality elements. Source: Reference Japan Science and Technology Federation (2021) added and modified (left), originally created by the author (right)

<Expected characteristics at the turning point>

- Transition from a period of economic growth to a period of social growth.
- An era in which physical fulfillment no longer responds to customer satisfaction.
- Manifestation period of products (products and services) of indifferent quality or reverse quality.
- Period of shift from consumption of goods to consumption of experiences.

Under the circumstances in which such characteristics appear, it means that true customer satisfaction cannot be expressed only by one axis of "physical fulfillment." Therefore, if the value of material consumption is reflected and the "customer satisfaction axis" is replaced by the "spiritual richness axis," which indicates individual values, it can be roughly illustrated as shown in Fig. 4.8.

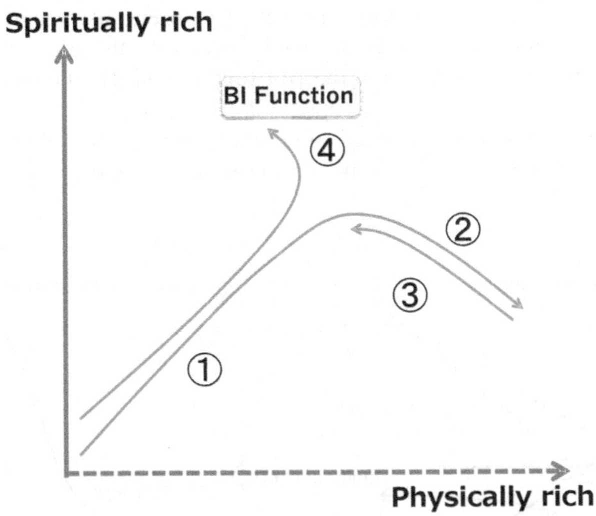

Fig. 4.8 Relationship between Kano Model and BI Function. Source: Reference (Sawaguchi et al., 2020).
① is a state in which physical abundance and spiritual affluence are satisfied, ② is a state in which convenience has caused an excess and the user can no longer be satisfied even though he/she becomes physically rich, and ③ is a means of restoring spiritual richness by going back in time (a kind of nostalgia). On the other hand, ④ shows a new direction that VE in the future should aim for, where spiritual richness is obtained by introducing BoI function (intentionally reduces physical richness). That is, ④ can be interpreted as the introduction of BoI function corresponding to the inverse quality suggested by Kano

4.3.2 *Future VE Covering BoI Function*

The main things mentioned so far are related and the three functions to be addressed in the future VE are organized in Fig. 4.9.

In fact, if the horizontal axis (*X*-axis) of the Kano Model (left and right graphs in Figs. 4.4 and 4.7) is the "inconvenience ↔ convenience" axis that indicates physical richness, and the vertical axis (*Y*-axis) is the "harm ↔ benefit" axis that indicates "true customer satisfaction" that also takes into account the mental richness of the user, it becomes possible to systematically organize the combinations of (inconvenience ↔ convenience) and (harm ↔ benefit) for each quadrant (see Fig. 4.10). That is, counter-clockwise, the first quadrant (benefit of convenience) is, corresponding to the second quadrant (benefit of inconvenience), the third quadrant (harmful inconvenience), and the fourth quadrant (harmful convenience).

As is clear from this figure, conventional VE has been to reach the benefit of convenience in the first quadrant from the harmful inconvenience in the third quadrant. As for future VE, "(1) activities to turn harmful convenience into the benefit of inconvenience" and "(2) activities to realize BoI functions from a zero base" can be assumed. Although the functional analysis approach in VE does not necessarily be used (see Fig. 4.1), "activities to turn the benefit of convenience into the benefit of inconvenience" and "activities to turn the harmful inconvenience into the benefit of inconvenience" are also naturally considerable.

Now, the characteristics of these four activities are summarized as follows:

Functions addressed in VE	Characteristics of each function		Natures of each function	Examples of definition of function (shaver)
Use function (the 1st function)	Tangible material consumption (In the case of products)	Practical function	Function which should be equipped with to use a product or service	Cut each hair of beard Prevent spread of shaved hair Absorb shaving impact etc.
Attractive function (the 2nd function)		Sensitive function	Function which makes the user own a product, such as design, appearance, service, slogan, etc.	Refreshing color Modern shape Tight-fitting texture etc.
BI Function (the 3rd function)	Functions of intangible experiential consumption (Mainly for individuals)		Function which creates awareness in spiritual fulfillment caused by inconvenient state of a product or service	(a part of it must be worked manually) Enables to improve shaving skills Gain a sense of safety and reliability etc.

Fig. 4.9 Characteristics of the three functions

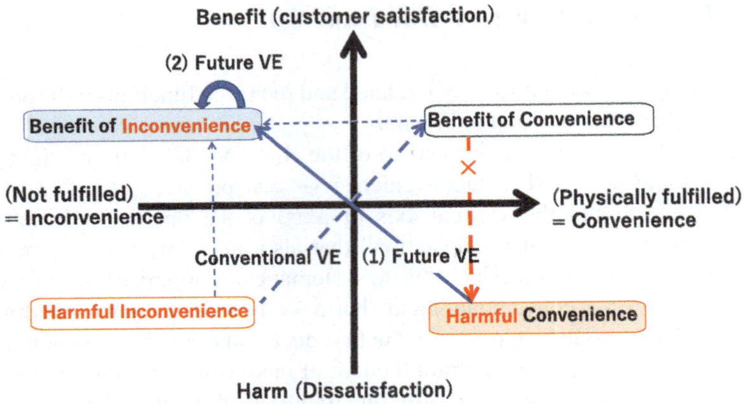

Fig. 4.10 Future VE which achieves BI functions

[Activities to Turn Harmful Convenience into the Benefit of Inconvenience (1) Future VE]
It is assumed that the user's sense of achievement has been greatly weakened for products that have fallen into a state of harmful convenience (EX. fully automatic EVs of the near future). At this stage, for example, by manualizing or deleting some of the useful functions which support fully automatic EV and realizing certain inconvenient functions, level of the "true customer satisfaction = benefit" of the target user could be enhanced. The author leaves to other books and papers (Kawakami, 2011; Matsuzawa, 2019)· for specific examples.

[Activities to Introduce BoI Functions at the Development Stage (2) Future VE]
It corresponds to VE at the planning and development stages, and because it is inconvenient from the beginning, it becomes a VE activity aimed at achieving the BoI function with the intention to obtain a certain benefit. Here, the use of the "Basic System Diagram of Inconvenient Function" (Matsuzawa, 2019) proposed by Matsuzawa et al. can be referred.

Figure 4.11 shows a functional diagram of "Foggy car navigation system" studied by "BoI & VE Study Group[Note)]." As shown here, the functional diagram is illustrated based on "the logic of purpose and means" organizing the functions from the right side (lower-order functions) toward the left side (higher-order functions).

<Functional diagram of ordinary navigation system (functional analysis)>

The upper side of Fig. 4.11 shows a functional diagram of a normal navigation system (composed of use functions). Details such as the procedure of making the diagram are omitted, but it has been clarified that the navigation system is a device for increasing movement efficiency.

<Functional diagram of foggy navigation system (functional analysis)>

The functional diagram of the foggy navigation system is illustrated on the underside of Fig. 4.11. Since the harmful convenience of "deteriorating the cognitive motivation of humans" has been confirmed, in order to solve this problem, a basic

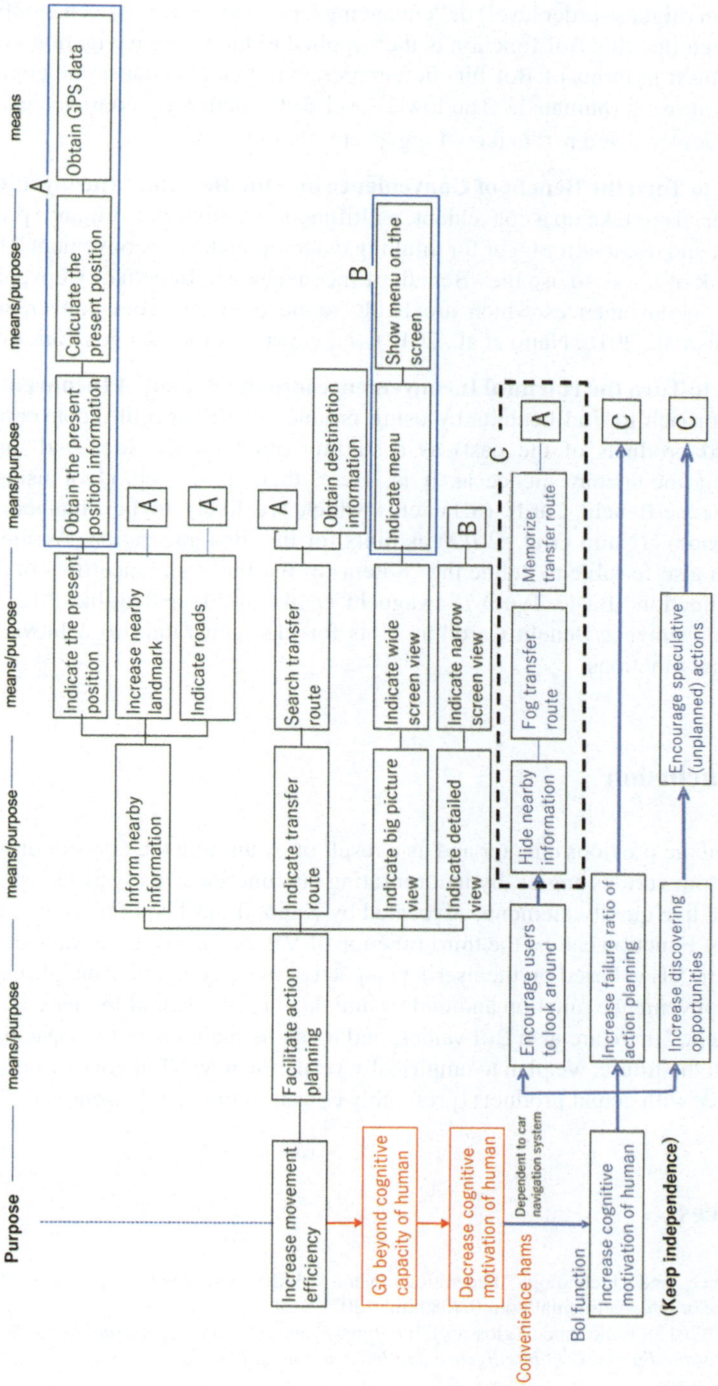

Fig. 4.11 Functional system diagram of navigation with BI functions added. Reference (Kawakami et al., 2016)

BoI function (highest-order level) of "enhancing independence" has been identified. It can be seen that this BoI function is then applied to the foggy navigation system by redefining it in terms of BoI function expression, it can "enhance the cognitive motivation of users (humans)." The lower-level BoI function group to achieve this BoI function is realized by "make ~ foggy," etc. (C in Fig. 4.11).

[**Activities to Turn the Benefit of Convenience into the Benefit of Inconvenience**] The approach is to take up a convenient, multifunctional, high-performance product as a subject and use it as a trigger for thinking to "try to make it inconvenient." Here, we can think of a way to use the "Benefit of Inconvenience/Benefit of Convenience Card" (12 inconveniences which are likely to be obtained from convenience)" (Kawakami et al., 2016; Naito et al., 2013) as a viewpoint for idea generation.

[**Activities to Turn the Harmful Inconvenience into the Benefit of Inconvenience**] It is an approach to find benefits by using products with harmful inconveniences (mainly old products of the past) as a starting point for the idea and making the most of the inconvenience as it is. Here, there is a method of using the "Inconvenience/Benefit Card" (8 benefits which are likely to be obtained from inconvenience)" (Naito et al., 2013) as hints for the idea generation. As the case of VE, it is also feasible to define the "8 items of the Inconvenience/Benefit Card" as "BoI Function (Basic Type) (Sawaguchi et al., 2020) and utilize "12 items of the Inconvenience/Benefit Card" as hints for idea generation to achieve these inconvenient functions.

4.4 Conclusion

Throughout the previous chapter and as a result of examining the concept of inconvenience from various angles, by incorporating the functional analysis viewpoint in VE and the five quality elements advocated by Kano, it has been found once again that the BoI Function can be the third function of VE. Also, since the value concept addressed in VE is based on the user's perspective (user-centered principle), if you add an inconvenience function and understand that "user's benefit increase = value improvement," there are also BoI values, and it can be included in the basic pattern of value. In the future, we plan to empirically verify the new VE theory summarized in this paper with actual products (preferably consumption of experiences).

References

Japan Science and Technology Federation: Kano Model and product planning, https://www.juse.or.jp/departmental/point02/08.html, (2021.8.29).

Kamiesu, H. (2015). Kano model (glossary), intelligence and information. *Journal of the Japanese Society for the Fuzziness of Intelligence and Information, 27*(4), 128.

Kawakami H (2011). Design born from inconvenience. Kagaku-Dojin Publishing Company, INC.
Kawakami, K., Sawaguchi, M., Matsuzawa, I., & Xiaolei, H. (2016). Value engineering introducing the value of inconvenience. *Value Engineering, 294*, 10–15.
Matsuzawa, I. (2019). Proposal of value creation method focusing on the third function "inconvenience". *Value Engineering, 305*, 9–15.
Murakami, M., & Watanabe, T. (2019). *Introduction to SDGs*. Nikkei Inc.
Naito, K., Kawakami, H., & Hiraoka, T. (2013). Design support method for utilizing the utility of inconvenience based on inventive problem-solving theory TRIZ. *Proceedings of the Society of Instrument and Control Engineers, 49*(6), 596–601.
Sawaguchi, M. (2015). *Introduction to Japanese-style manufacturing engineering*. Doyukan.
Sawaguchi, M. (2020). *The first step of planning and development method*. Doyukan.
Sawaguchi, M., Kawakami, H., Matsuzawa, I., Miyata, R., & Nishiyama, K. (2020). *Emanuel LELEITO: Implementation of benefit of inconvenience*. Modern Science Digital.
Tsuchiya, H. (1982). *VE ~ improvement Technology for Cost Reduction*. Japan HR Association.
Yoneyama, T. (1969). *Quality control manual*. Japan Science and Technology Federation.

Manabu Sawaguchi Graduated from the Department of Mathematical Engineering, Faculty of Engineering, Keio University in 1982. Completed the doctoral program at the Graduate School of Science and Engineering, Waseda University in 2005. Doctor of Engineering. Engaged in research on value creation engineering utilizing management technology (mainly VE, TRIZ, QFD, etc.). After serving as a professor at the Department of Management Design, Graduate School of Science and Engineering, Waseda University since 2010, he became a professor at the Graduate School of Technology Management, Ritsumeikan University in 2019. Counselor of Society of Japanese Value Engineering, Vice President of Japan TRIZ Association, Director of the Japan Society of System Design, Member of the Japan Society of Industrial Engineers, Japan Society of Management Systems, Member of the Japan Society of MOT Studies. In 2016, he received the VE Academic Achievement Award (Society of Japanese Value Engineering). And he received Academic Award (Japan Association for Management Systems) in 2022.

Note: "BoI & VE Study Group" (Value Design Lab) The author (Sawaguchi), who also serves as the director of the Value Design Laboratory of Society of Japanese Value Engineering, became the project general manager, and with the participation of Professor Koji Kawakami (currently Professor of Kyoto University of Advanced Sciences), a leading expert in the research of the Benefit of Inconvenience, the study group conducted research activities from October 2015 to 2020. Other study group members are: Dr. Nishiyama (Nagoya Univ.), Dr. Emanuel (Nagoya Univ.), Mr. Matsuzawa (IHI), Ms. Miyata (IHI), Secretariat: Ms. Ono (SJVE)

Chapter 5
Supporting Human Activities by Obstruction

Kazushi Nishimoto

Abstract This chapter describes "Support by Obstruction " technologies that the author's laboratory has been studying. Support by Obstruction refers to introducing an obstructing element that does not originally exist in an action, thereby producing a positive effect that facilitates, streamlines, or promotes the performance of the action itself, or some or all its peripheral actions. This concept is useful to create contrivances for everyday learning by embedding physical or intellectual hurdles to be overcome in our daily activities. In many cases, introducing obstructing elements causes inconvenience. Therefore, this concept seems similar to the concept of the "Benefit of Inconvenience." Indeed, there are cases that can be regarded as both a case of Benefit of Inconvenience and a case of Support by Obstruction. However, there are essential differences between them. This chapter firstly defines the Support by Obstruction concept, shows the relation with the benefit of inconvenience, illustrates several examples of applications based on the concept of Support by Obstruction, and finally describes a future direction of research on Support by Obstruction and its applications.

Keywords Support by obstruction · Maintaining public good · Skill acquisition support · Training support · Everyday (re)learning

5.1 Introduction

5.1.1 Definition of "Support by Obstruction"

The author's laboratory is promoting research on "Support by Obstruction" technologies. Support by Obstruction (SbO) refers to introducing an obstructing element

K. Nishimoto (✉)
The Graduate School of Advanced Science and Technology, Japan Advanced Institute of Science and Technology, Nomi, Ishikawa, Japan
e-mail: knishi@jaist.ac.jp

H. Kawakami (ed.), *Systems Design Based on the Benefits of Inconvenience*,
Translational Systems Sciences 31, https://doi.org/10.1007/978-981-19-9588-0_5

that does not originally exist in an action, thereby producing a positive effect that facilitates, streamlines, or promotes the performance of the action itself, or some or all its peripheral actions. In some cases, an obstructing element is introduced by adding a new function for the disturbance, while in other cases, an obstructing factor is created by reducing a useful function that already exists. Although there are a variety of possibilities for the target activities of support, we are currently focusing mainly on activities that improve a person's intellectual and physical abilities and aim to embed these activities in everyday life in as natural a manner as possible.

In ordinary engineering thinking, the common means of supporting something is to remove or reduce the obstacles that prevent it from being done, so supporting something by obstructing it may intuitively seem like a contradiction in terms. However, it is quite common to have a positive effect on something by obstructing it. For example, a typical example is a car stop pole at the entrance of a park. By obstructing cars from entering the park, they support the safe activities of the park's inhabitants. A more familiar example is the password lock on house keys and computers, which can also be regarded as a kind of SbO. Thus, there are many examples of the use of obstructive elements to ensure people's safety, maintain security, and ensure the public good. Many of them fall into the pattern of what D.A. Norman calls "lockout" (Norman, 1988).

5.1.2 Relation between "Benefit of Inconvenience" and "Support by Obstruction"

Some may wonder if the case of the car stop pole can be also a case of "Benefit of Inconvenience (BoI)," and therefore BoI and SbO are just a paraphrase of the same thing. It is often the case that the same case is both a case of BoI and a case of SbO. However, there are many cases of BoI but not of SbO, and vice versa.

The car stop pole does not fall under the category of a BoI case. This is because of the principle that a case in which "the inconvenience of others is a benefit to oneself" is not recognized as a BoI case (Kawakami, 2013). The definition of BoI states that the person who suffers the inconvenience and the person who enjoys the benefit must be the same. In the case of the car stop pole, the person who benefits (= pedestrian) and the person who suffers inconvenience (= driver of the car) do not coincide, so this case is not a BoI. On the other hand, SbO does not exclude the case where "another person's inconvenience is my benefit." This is one of the points where SbO differs from BoI.

Conversely, an example of a BoI case, but not a SbO case, is a washboard. Compared to a fully automatic electric washing machine, washing clothes using a washboard requires more time and effort. In this respect, the washboard is considerably more inconvenient than the washing machine. However, the structure of the washboard makes it possible to perform such tasks as carefully removing small stains and fine-tuning the washing process, such as washing only certain parts of the garment gently and scrubbing other parts, which are not possible with a washing machine. In this respect, the washboard is beneficial, although it is time consuming. Therefore, the washboard can be regarded as a case of BoI. However, a washboard is not a thing made by adding obstructive elements to a washing machine, nor is it a thing made by removing some of the useful features of a washing machine. Even without comparing it to a washing machine, a washboard was not created by adding any obstructive elements to something in the first place. Therefore, a washboard cannot be considered a case of SbO.

By the way, as in the case of the washing machine and washboard mentioned above, when new and more convenient things are produced because of technological progress, it often happens that the things of the past, which were considered convenient until then, become relatively inconvenient. However, when the characteristics or functions that the past things possessed are lost in the new things, these lost features are sometimes recognized as a BoI case. Such a thing in the past was not designed based on the concept of BoI, but the perception of the thing changed from a "convenient thing" to an "inconvenient thing" due to changes in value criteria with the transition of time and technology, even though the thing itself has not changed in any way. Thus, the concept of BoI has an epistemological aspect. On the other hand, SbO is a design-theoretic concept for designing a secondary benefit by adding a disturbance factor to a certain thing and is not easily affected by changes or differences in people's perceptions.

Thus, although BoI and SbO have some parts in common, they are not merely the same thing rephrased from the "result-side perspective (i.e., inconvenience)" and the "cause-side perspective (i.e., obstruction). They are fundamentally different concepts.

5.2 Assist in Maintaining the Public Good by Obstructing

As described in the previous chapter, the conventional target of SbO was the area of safety, security, maintenance of public good, etc., and the element of obstruction was used to discriminate between those who have the right to use a certain thing and those who do not. Here, we introduce a case in which the maintenance of public good is also a target of support, but instead of discriminating based on the existence of rights, we take measures to interfere with the exercise of rights when obligations accompanying the exercise of rights are not fulfilled, to demand the fulfillment of obligations.

TableCross (Nishimoto et al., 2011) is a system designed to encourage users of a shared communication space in a laboratory to maintain and manage the space. The upper photo in Fig. 5.1 shows a shared communication space in the author's laboratory. The space is actively used as a place where various talks and discussions can be held freely. Since eating and drinking are allowed, many visitors bring their drinks and food. However, there is no end to the number of people who leave empty drink bottles, food wrappers, and other such items behind after use without cleaning them up.

To solve this situation, we created the TableCross system. We placed a retroreflective cloth as a tablecloth on a table in a shared communication space and installed an infrared light source and an infrared camera above the table. The light from the infrared light source is reflected by the retroreflective cloth on the table and captured by the infrared camera. If an object is placed on the table, the infrared light is not reflected from that part of the table, and it becomes a shadow on the captured image. The resulting infrared image of the tabletop is binarized (center of Fig. 5.1) to determine the ratio of the shadow to the total surface area of the tabletop, which is the degree of clutter on the table. For example, if the obtained degree of clutter is 70%, the desktop screen of the personal computer used by each user of the shared communication space for personal work generates "garbage icons" that fill 70% of the area and scatters them on the desktop screen (Fig. 5.1, bottom). Even if the garbage icons are deleted by an operation on the PC, they soon resurface and are scattered on the screen unless the table in the shared space is reorganized. This system is intended to stimulate users' awareness of the need to maintain and manage

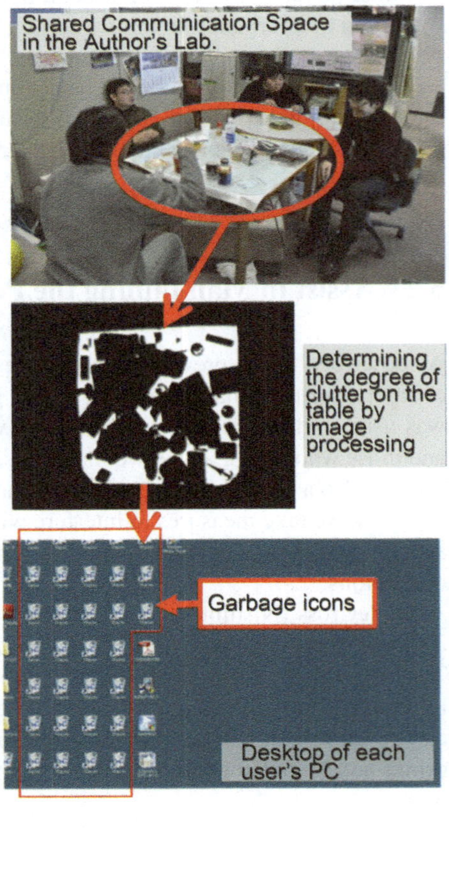

Fig. 5.1 TableCross system

the shared space by making their personal space dirty when they defile the shared communication space.

The scattering of garbage icons on PC desktops in TableCross is an obstructive function. However, this obstructive function is expected to have the utility of ensuring the public good of maintaining the shared space.

5.3 Assistance in Gaining Competence by Obstructing

One of the areas of application where SbO is particularly effective is in the acquisition of skills, such as learning and training. In learning and training, learners are often given tasks that are a step higher than their current abilities and forced to work harder to overcome them to acquire new knowledge and abilities. These tasks are often compared to hurdles and can be seen as a kind of obstructive element. However, in normal learning and training, where the main goal is to reach a higher level than the current level of ability, it is not appropriate to consider the task as an obstacle in this case because overcoming the task is itself an action to be performed. For the task to be considered an obstructing factor, it must be a situation in which there would not normally be such a task to overcome. In this section, two such situations will be discussed and examples of research on each will be presented.

5.3.1 Assistance in Acquiring Skills in Situations where Learning or Training Is Not the Primary Objective

The Gestalt Imprinting Method (G-IM) (Nishimoto & Wei, 2015) is a Chinese character input system with a function to prevent forgetting Chinese characters. In recent years, with the spread of personal computers and smartphones, opportunities to write Chinese characters by hand have become extremely rare. As a result, the number of "Character Amnesia" people in Japan and China, who can read but cannot write Chinese characters, is said to have increased rapidly and has become a social problem. This is thought to be because currently popular Chinese character input systems, both in China and in Japan, use a method whereby pronunciation is input and converted into Chinese characters. In this system, when the system outputs the Chinese character corresponding to the input pronunciation "sai," for example, the user checks whether the purposed character "歳(sai)" is not converted to a homophone such as "再(sai)," but if the correct character "歳(sai)" is output, the user does not check the details of the character. In this way, the lack of attention to the details of the character forms when inputting Chinese characters during the creation of a document can cause the user to gradually forget the correct forms of the characters and become unable to write them.

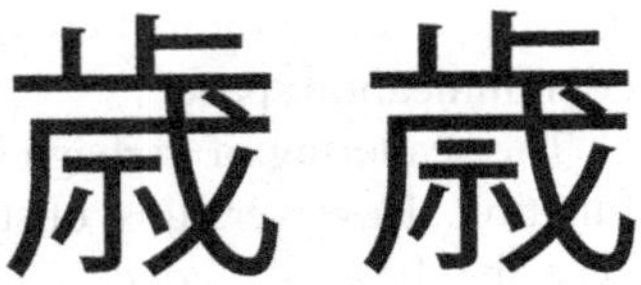

Fig. 5.2 Examples of correct (left) and incorrect (right) Chinese characters output by G-IM

To solve this problem, G-IM employs an obstructive measure: G-IM sometimes outputs a Chinese character with an incorrect shape, and the document cannot be saved unless it is corrected to the correct shape. Figure 5.2 shows an example of a Chinese character with the wrong shape and its correct shape. If G-IM outputs a Chinese character with the wrong shape (for example, the right side of Fig. 5.2), it will be replaced with the correct shape (the left side of Fig. 5.2) when the user selects the character and performs the reconversion operation. If this correction is not made, the document cannot be saved. Therefore, while writing a document using G-IM, the user is always forced to pay attention to the detailed shapes of the Chinese characters. Furthermore, by correcting misshaped characters, the system will always present the "correct form of the characters." We conducted an experiment comparing G-IM with a conventional Chinese character input system that converts pronunciations into Chinese characters, and with handwriting. As a result, we found that G-IM significantly strengthened the shape memory of Chinese characters compared to the other methods. The results showed that G-IM significantly enhanced Chinese character shape memory in comparison with the others. The reason why handwriting does not enhance character shape memory is that handwriting does not correct incorrectly written characters and does not provide an opportunity to learn the correct shape of the characters.

In this way, when a task for acquiring some kind of ability is embedded in an activity that has a different main objective from learning or training, and the task is to overcome the embedded task while acting to achieve the main objective, the task becomes an unnecessary obstacle to the main objective at the same time. The embedded task is an unnecessary obstacle from the viewpoint of the main objective. However, the obstacle can support the secondary objectives of learning and training at the same time as the main objective.

5.3.2 Training Support for Already Acquired Competencies and Knowledge

Apollon 13 (Yokoyama & Nishimoto, 2010) is a system that supports training to avoid performance stoppages due to unexpected events such as mistakes in live situations such as music recitals and concerts. When attending a performance, performers practice well in advance and acquire the ability to play the piece they are going to perform completely, so they only need to demonstrate their acquired ability in the actual performance. However, during the performance, the performer may make mistakes that he or she would never normally make due to nervousness or other factors, and as a result, the performance may come to a halt in the worst-case scenario. To avoid such a situation, it is necessary to conduct training in advance to deal with such mistakes. However, since performers have already acquired the ability to play the piece they are performing, there is no way to train them, since mistakes rarely occur during the rehearsal period. Therefore, Apollon 13 incorporates a performance jamming function, which artificially generates a performance error state by occasionally replacing the playing note with an adjacent note or other incorrect note, even though the player is playing correctly. By practicing with Apollon 13, the player can acquire the ability to deal with performance errors in a live performance.

In Apollon 13, unlike the G-IM presented in the previous section, the object of the SbO element is the main goal itself, which is musical performance. In this sense, it has the same structure as in the case of ordinary learning or training, where overcoming a task is itself an action to be performed, as pointed out at the beginning of Sect. 5.3. The difference between Apollon 13 and ordinary learning and training is whether or not the trainee has completed acquiring all the skills and knowledge to be trained. In the case of ordinary learning and training, the trainee has not yet acquired the skills and knowledge to be trained, so the task is given to the trainee as a task that has not yet been mastered. On the other hand, if the trainee has already mastered the skill or knowledge, there are no tasks left to be given to the trainee. Therefore, in the case of a musical performance, the only thing that can be done is to just repeat the entire performance before the actual performance. Apollon 13, therefore, makes it possible to conduct emergency training and ability maintenance training for acquired skills and knowledge by creating virtual situations that would not normally occur through the introduction of obstructing elements, such as the temporary loss of acquired skills and knowledge or the loss of the ability to demonstrate them accurately.

5.4 Future Direction of Support by Obstruction

5.4.1 Breaking Free from Inconvenience

G-IM, as exemplified in Sect. 5.3.1, adds a function for the secondary purpose of learning Chinese characters by introducing an obstructive element, the output of misspelled Chinese characters, into the main task of document creation. Because of the additional effort required to overcome this obstacle, G-IM is more inconvenient than ordinary Chinese character input systems. This can be seen as a case of BoI since the benefit of learning the correct form of a Kanji character is gained by this inconvenience.

However, we believe that the efforts to SbO should ultimately aim at moving away from a state that can be regarded as an inconvenience. This is because we often receive comments from people who have used G-IM, such as, "I understand the merit of learning correct Chinese characters, but I don't want to use G-IM for everyday document creation because it is still inconvenient." This indicates that the disadvantages of inconvenience outweigh the advantages of learning. For people to be able to enjoy the advantages without complaints, it is necessary to at least reverse this relationship.

Even when G-IM is used to create documents, its main purpose is document creation, not Chinese character learning. Nevertheless, G-IM introduces the task of constantly checking the detailed glyphs of all Chinese characters and correcting them as necessary, which is as much or more burdensome than when the main purpose is to learn Chinese characters. This is what is called "too high a hurdle," and as a result, many users may feel that they do not want to use the system. Therefore, even if learning Chinese characters is a secondary purpose, the same level of burden should not be introduced as in the case of regular Chinese character learning. It is necessary to devise new tasks that are qualitatively and quantitatively less demanding. Ideally, some kind of disturbance should be introduced so that the user is not aware of the disturbance, although he/she can certainly benefit from it. If this situation can be achieved, the user can enjoy the benefits without any inconvenience. In other words, SbO tends to be a BoI in many cases at the beginning, but we believe that it should ultimately aim for a benefit rather than an inconvenience.

5.4.2 *An Example of SbO without the Inconvenience*

An example of a situation in which the user is unable to recognize the existence of obstacles and only benefits from them is the iDAF drum (Nishimoto et al., 2014). This is a practice support system for a drum performance, especially for effective training of the extensor carpi radialis muscle (hereinafter referred to as "extensor muscle"), which is used to turn the wrist back toward the dorsal surface of the hand. When playing the drums, after striking the surface of the drum with the stick, the player must "actively" raise the stick while consciously using the extensor muscles to turn the wrist back. This improves the sharpness and sound quality of the strike and also enables high-speed continuous striking. However, it is very common for many players to passively raise the stick without using their extensor muscles, letting the repulsive force from the drumming surface do the work for them.

We have developed an extensor muscle training system for drum performance using insignificantly delayed auditory feedback (iDAF). Delayed auditory feedback (DAF) is generally defined as a delay of 100–200 ms in the speaker's speech that is fed back to the speaker's ear. This is known to cause stuttering-like phenomena in which speech is not smooth and sounds are repeated or stretched (Lee, 1950). iDAF is auditory feedback with a very short delay, where the delay is shortened to a level where the presence of the delay is not perceptible to the human auditory system. It is known that it takes about 30 msec for humans to begin to perceive a gap between an action and the resulting sound (Nishibori et al., 2003). iDAF drum is a system that

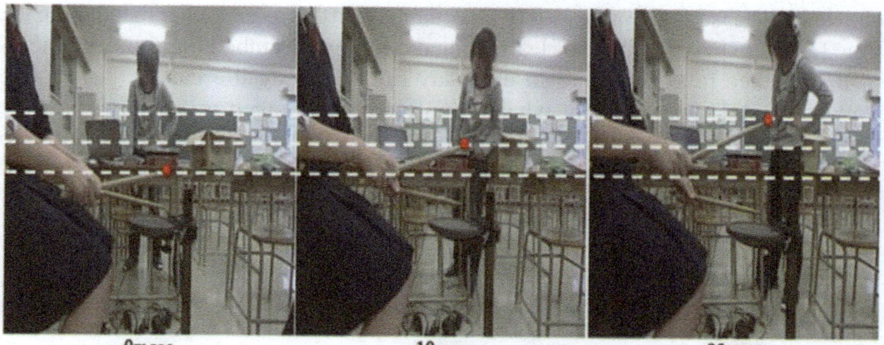

Fig. 5.3 Example of change in stick swing up position due to change in delay time

provides a delay of less than 30 ms (20 ms in the experiment) between the striking of a surface with a stick and the generation of a striking sound. When such a slight delay is applied, most players cannot feel the presence of the delay, and interviews with players revealed that they did not feel any change in their playing sensation with or without the slight delay. Nevertheless, when a delay is applied, there is a clear change in the amplitude of the stick. Figure 5.3 shows examples of changes in drumstick swing position under three conditions: no delay (0 ms), 10 ms delay, and 20 ms delay. Thus, by using slightly delayed auditory feedback, it is possible to make the subject swing the drumsticks up more widely without changing the drumming sensation. Experiments with subjects using the iDAF drum showed that the extensor carpi radialis muscle was better used during practice with the iDAF drum. In some cases, furthermore, the tone quality was improved.

The slightly delayed auditory feedback of the iDAF drum is an obstructing element, but it is not perceivable. However, this imperceptible disturbance can be used to effectively train the extensor carpi radialis muscle, which is important in drumming. In this sense, although the iDAF drum is an example of SbO, it should be an example of the "Benefit of Convenience" rather than the "Benefit of Inconvenience" because it does not cause any inconvenience.

5.5 Conclusion

In this chapter, I first illustrated the concept of Support by Obstruction (SbO), introduce examples of the application of SbO to the maintenance of public good and the acquisition of abilities, and showed the direction to move away from inconvenience. So far, our laboratory has studied systems that support various activities of people grounded on the idea of SbO. Based on our research experience, the idea of SbO is very compatible with the purpose of learning and training, as shown in the case study presented in this chapter. Furthermore, rather than taking the time for the express purpose of learning and training, by embedding small elements of disturbance here and there in everyday life, everyday life can be turned into a place of skill improvement where people are encouraged to gradually acquire knowledge and skills in their daily activities. We will conduct further research to achieve a convenient environment for intellectual and physical improvement based on the concept of Support by Obstruction.

References

Kawakami, H. (2013). Do Set Up Inconveniences. *Journal of Japanese Society for Artificial Intelligence, 28*(4), 615–620. (in Japanese).

Lee, B. S. (1950). Effects of delayed speech feedback. *Journal of the Acoustical Society of America, 22*(6), 824–826.

Nishibori, Y., Tada, Y., Sone, T. (2003). Study and experiment of recognition of the delay in musical performance with delay. IPSJ SIG Technical Report, 2003-MUS-53(9), pp.37–42 (in Japanese).

Nishimoto, K., Ikenoue, A., Shimizu, K., Tajima, T., Tanaka, Y., Baba, Y., Wang, X. (2011). Table-Cross: Exuding a shared space into personal spaces to encourage its voluntary maintenance. CHI2011 Extended Abstract, 1423–1428.

Nishimoto, K., Ikenoue, A., Unoki, M. (2014). iDAF-drum: Supporting everyday practice of drum by adding an unperceivable factor. Proc. 9th Int'l. Conf. on Knowledge, Information and Creativity Support Systems, pp.384–395.

Nishimoto, K., Wei, J. (2015). G-IM: An Input Method of Chinese Characters for Character Amnesia Prevention. Proc. The Eighth International Conference on Advances in Computer-Human Interactions (ACHI2015), pp.118–124.

Norman, D. A. (1988). *The psychology of everyday things*. Basic Books.

Yokoyama, Y., Nishimoto, K. (2010). Apollon13: A training system for emergency situations in a Piano performance. Active Media Technology, LNCS6335, Springer, pp.243–254.

Kazushi Nishimoto He received his MS in Mechanical Engineering from Kyoto University in 1987 and joined Matsushita Electric Industrial Co., Ltd. In 1992, he joined ATR Communication Systems Research Laboratories as a researcher and became a visiting researcher at ATR Media Integration & Communications Research Laboratories in 1995. In 1999, he joined the Japan Advanced Institute of Science and Technology (JAIST) as an associate professor and became a full professor in 2007. He received the 1999 Sakai Memorial Award from Information Processing Society of Japan (IPSJ), the 1999 Best Paper Award from the Japanese Society for Artificial Intelligence, the ACM Multimedia 2004 Best Paper Award, the 14th Human Interface Society Best Paper Award, and others. He is a member of the IEEE computer society, ACM, Information Processing Society of Japan, Japanese Society for Artificial Intelligence, and Japan Society of Kansei Engineering. Dr.Eng. Fellow of the Information Processing Society of Japan. He engaged in research on support technologies for creative activities, support technologies for intellectual activities by the obstruction, and technologies for utilizing unused knowledge.

Chapter 6
Weak Robots: Relational-Oriented Approach to Human Well-being

Michio Okada

Abstract Many robotics researchers have attempted to design robots that can move around on their own, believing that autonomous robots should not need help from social others. However, even high-performance robots and their artificial intelligence technologies still have many weaknesses and imperfections. Rather than claiming to be able to do this or that, the robot should recognize its own weaknesses and imperfections and ask others to help it with what it cannot do.

This chapter introduces the concept of "weak robots," i.e., robots that are somewhat imperfect and require the help of other people. For example, the "Sociable Trash Box" robot cannot pick up trash by itself; however, it can pick up trash with the help of children. In addition, the "iBones" robot stands on a street corner and hesitantly attempts to give tissues to passersby. The "Talking-Bones" robot attempts to tell an old story to a child; however, it sometimes forgets important words. All of these robots require assistance and are somewhat inconvenient, however, the weak robots motivated the positive involvement of the children, bringing about the children's strengths and ingenuity. In addition, the expressions of the children were cheerful and somewhat satisfied.

It can be said that these will go beyond the conventional values of providing convenience with highly functional robots and realize a state of well-being for the children, i.e., their abilities are fully utilized in cooperation with the weak robots, and they feel a sense of happiness.

Keywords Weak robot · Human–robot interaction · Social robot · Sociable trash box · Well-being

M. Okada (✉)
Toyohashi University of Technology, Toyohashi, Japan
e-mail: okada@tut.jp

H. Kawakami (ed.), *Systems Design Based on the Benefits of Inconvenience*,
Translational Systems Sciences 31, https://doi.org/10.1007/978-981-19-9588-0_6

6.1 Introduction

Robots that once only appeared in cartoons and works of science fiction can now be found on street corners, and cleaning robots are found in many homes. In addition, fully automated, self-driving systems will become a reality in the near future.

However, if we think about it dispassionately, robots and artificial intelligence systems have many weaknesses and imperfections (similar to humans). Why not just expose those weaknesses in moderation without hiding it?

From the perspective of benefits of inconvenience research, this chapter introduces the concept of "weak robots," i.e., robots that are somewhat imperfect and require the help of other people (Okada, 2012, 2017, 2022). For example, the "Sociable Trash Box" robot (Yamaji et al., 2011) cannot pick up trash by itself; however, it can pick up trash with the help of children. In addition, the "iBones" robot stands on a street corner and hesitantly attempts to give tissues to passersby. The "Talking-Bones" robot attempts to tell an old story to a child; however, it sometimes forgets important words. All of these robots are somewhat imperfect and require assistance. The aim of this study is to sort out the social significance of these weak robots.

Until recently, the research and development of autonomous robots have been dominated by the individual capability-based approach, which attempts to make functions and performances self-contained within the body of individual robots. Here, design by addition has been pursued by adding various functions from the viewpoint of how to fill the space in the robot's capabilities and functions. In our research, we have focused on weak robots and how we can design relationships with other people based on the premise that robots are inherently imperfect.

Thus, based on the weak robot concept, we consider peaceful coexistence between humans and robots wherein the strengths and weaknesses of both humans and robots are complemented and supported.

6.2 Weak Robots

6.2.1 Origin of Weak Robots

In our laboratory (ICD-LAB), many types of social robots have been proposed and created as tools for our daily communication research. We were originally working in the spoken language processing and cognitive sciences in communication fields; thus, we were a complete novice when it came to robotics and related technologies. With a lot of hard work, a simple sociable creature called "Muu" (Okada, 2000) was created. With the hope that someday this robot would be able to take care of children, we brought Muu into a kindergarten class.

We attempted to have the children and the Muu robot play with building blocks. The children could play with the building blocks alone or while interacting with the

Fig. 6.1 Sociable creature "Muu" and a child playing with blocks

Muu robot (Fig. 6.1). Once the children became accustomed to this robot, we tried to have Muu give instructions to the children, e.g., "Put in the flat yellow building block next!" Here, we wanted to observe the extent to which the children would manipulate the building blocks in response to requests from the Muu robot.

Eventually, the children became tired of Muu's clumsy physical behaviors and began to care for the robot, asking questions like, "What should we do next?" prior to the robot's instructions. The idea of robots caring for children is an empty theory on our desk. In reality, the children's abilities were overwhelmingly superior to those of the robot, and the children were incredibly excited while caring for the robot.

While observing the children interact with the robot, we began to consider if there is positive meaning and value in the weakness of robots.

6.2.2 Weak Robot Concept

We appear to have developed a culture that believes performing tasks independently is a good thing. Even when raising our children, we sometimes want children to learn to put on their socks by themselves as soon as possible. In response, children may sometimes boast, "I can already do it by myself, isn't that great?" Even in formal education, there is an unwritten rule that tests must be taken independently; we cannot rely on anyone else's help and students support their efforts by believing that "I can solve this problem by myself."

This is similar in the autonomous robotics field. Assuming that an autonomous robot should not rely on the help of others, we attempted to design a robot that can move around by itself. However, there are many weaknesses and imperfections in high-performance robots and artificial intelligence technologies. Rather than claiming to be able to do this or that, why do not robots recognize their own weaknesses and imperfections and ask those around them to help them with what they cannot do?

Based on such thoughts, the concept of a "robot that is not complete by itself, but rather is half-receptive to others, skillfully draws others out, and works together to achieve a goal" has been called a "human-dependent social robot" or a weak robot. Weak robots appear somewhat imperfect, and they are typically perceived as cute and make us want to support them. In the following, we introduce representative weak robots constructed in our laboratory.

6.2.3 Sociable Trash Box

When planning a new robot in our lab, it is easy to think about it in terms of a "robot that does XX." We hope that useful robots will enrich our lives, however, what would life be like with only such convenient robots around us? Is there any room left for our own participation? What if a robot required a little more work? From such questions emerged the concept of the Sociable Trash Box robot, which is unable to pick up trash by itself and elicits the help of children to perform this task. This concept originated from the casual relationship between infants and their caregivers.

As "weak beings" who are held in their caregivers' arms and cannot do anything independently, infants can struggle to get the milk they need, and they can move around as they wish while using positive interpretations of those around them. Sometimes, a toddler, unable to do anything on his own, can take away his siblings' favorite toys while keeping those around him on his side. In that sense, they are also the "strongest being" in the home.

This can also be regarded as the initial emergence of social skills required by infants to survive. In contrast to an action strategy based on individual competences, where individuals attempt to complete all tasks by themselves, this is referred to as a relational-oriented action strategy whereby individuals achieve their objective while eliciting help from social others. In other words, this action strategy is a skill that social robots have yet to obtain.

Therefore, I decided to create the Sociable Trash Box robot, which only involves a small laundry basket and motorized wheels, and I brought this robot to a playground where children were playing (Yamaji et al., 2011).

The children who noticed the somewhat strange-looking trash box immediately gathered around, asking, "What is it?" (Fig. 6.2). Perhaps getting a sense of feelings of the robot, one child threw the paper bag in their hand into the Sociable Trash Box. The robot responded with a bow-like gesture. It is unknown whether this was a gesture of gratitude for helping or a request for more help. Encouraged by this gesture, the nearby children began to search for garbage, and the storage space of the robot was filled with garbage. This somewhat unreliable robot, which cannot pick up trash by itself, was able to achieve its goal of picking up trash by interacting with the children.

Fig. 6.2 Sociable Trash Box picking up trash with children

As another unexpected aspect, the children also began to sort the garbage. Perhaps feeling bad that different types of trash were being thrown into the Sociable Trash Box haphazardly, the children began to assign roles to robots with different colors, e.g., plastic bottles for the red robot and paper for the gray robot.

Had the Sociable Trash Box robot picked up the garbage by itself, the children would not have interacted with it. In that sense, the weaknesses and imperfections of the Sociable Trash Box motivated the positive involvement of the children, bringing about their strengths and ingenuity. In addition, the expressions of the children were cheerful and somewhat satisfied. For example, one child said, "I don't feel bad about helping a robot."

According to the cognitive scientist Yutaka Saeki (Saeki, 2017), everyone has the need to care for someone else and this is not striving for a better life for oneself but rather striving for a better life for someone other than oneself and ultimately living a better life.

In this context, the "someone" is not necessarily a person. In our daily lives, we frequently take care of plants, dogs, or cats. In addition, children may care for a wide range of things, from ants or other insects to small animals or even dolls. In the initial phase of our research on weak robots, we were familiar with the concept that everyone has a need to care for someone else. We feel happy when we give or receive help and achieve goals together.

6.2.4 iBones

Another example of a weak robot is our iBones robot, which attempts to hand out pocket tissues to passersby on street corners. Handing out pocket tissues is a common guerrilla marketing method in Japan.

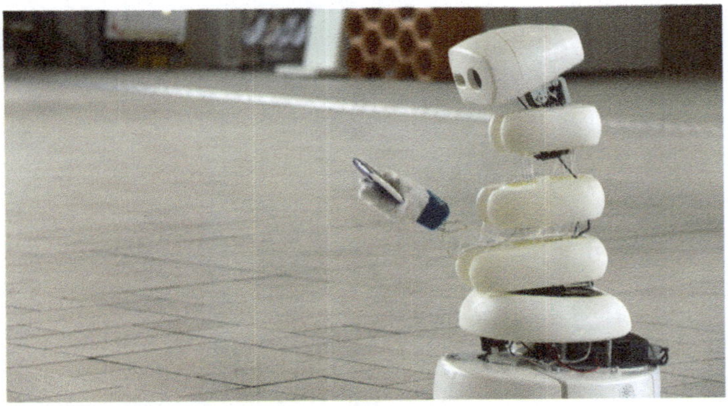

Fig. 6.3 iBones handing out pocket tissues

Here, rather than developing a convenient robot that hands out tissues at street corners, the goal was to develop a robot that allows constructive exploration of social interactions.

We believe that social interactions between humans and robots are extremely interesting. With the iBones robot, pocket tissues are handed to a stranger who happens to be walking by, and the robot is an unknown presence to many people. When attempting to hand out the tissues, if the person does not accept them, the tissue cannot be handed out. Similarly, when attempting to receive the tissue, if the robot does not hand them out properly, the tissue cannot be accepted. Both parties must partially trust the other with an action, and the exchange will fail if both sides do not share the same goal.

Figure 6.3 shows the iBones robot handing out tissues. The iBones robot uses information acquired using a camera to identify individuals and follows their movements. When a person comes near the robot, the robot attempts to hand the person a tissue. However, human movements are unique and agile; thus, it is difficult to match the timing. Each time a person appears in front of the robot, when the robot attempts to hand out the tissue, if it determines that the action will fail, it pulls back in a disappointing manner and repeats the action. This results in the appearance of somewhat fidgety behavior.

Perhaps feeling bad for such a robot, we observed an older woman approach the robot and stop. She happily accepted a tissue while coordinating her timing with the iBones robot's hand movements. Note that this represented a more skillful reception of the tissue by the woman than skillful handing out of the tissue by the robot.

When handing out tissues, there is a brief moment at which the two sides share the same mindset and share a common goal, and they adjust the timing of their movements to realize that goal. One reason why the older woman appeared to be happy may have been a combination of the sense of accomplishment for achieving the successful exchange and a sense of connection with the robot. As demonstrated by the relationship between the Sociable Trash Box and the children, there appears to be a state of well-being (Calvo & Peters, 2014) in which one's abilities are utilized to achieve an active and happy state.

At first glance, the hesitation and awkwardness of iBones may seem inefficient; however, these characteristics appear to motivate the cooperative nature of social others.

6.3 Weak Robot Concept in the Field of Communications

Here, I introduce the benefits and values of spoken utterances with ambiguous or incomplete meanings in our communication with robots.

The Sociable Trash Box attempts to bend its upper body when it detects that a child has thrown trash into the storage space. This gesture can be interpreted in various ways, e.g., "thank you for throwing away trash here!" or "pick up more trash!" Some may think this alone sends an ambiguous message, which may be perceived by some as an inconvenient communication style. However, leaving room for such interpretations is considered an important design element when drawing out positive interpretations of the surrounding people or their participation.

For example, we could implement a synthetic voice system in the Sociable Trash Box and make it speak in Japanese. If the words spoken were "thank you very much," it could be clearly interpreted as a gesture of gratitude. However, a repeated call of "I found trash. I found trash. Please pick up this trash!" would likely be somewhat annoying. In addition, it would feel like the robot is unilaterally giving instructions or commands, which is likely to make people uncomfortable.

According to the literary critic Mikhail Bakhtin (Bakhtin, 1996), an utterance with a self-contained meaning is referred to as authoritative discourse and tends to impose its meaning unilaterally without room for coordination with the listener. The utterance from the robot "Please pick up this trash!" somehow sounds harsh and strong.

In addition, the interpretation of incomplete phrases whose meaning is not self-contained is partially entrusted to the listener, and meaning is formed in a mutual manner. Here, there is room for adjustment with and acceptance from the listener, ultimately resulting in persuasive power.

6.3.1 Moco Language with Sociable Trash Box

With the recent Sociable Trash Box, we are attempting to use semi-syllabic sounds (i.e., a "Moco language") that are not yet articulated sufficiently, e.g., "Moco" or "Mocomon!" Assume a scenario where the Sociable Trash Box walks around a plaza endlessly humming, "Mocomon, Mocomon!"

Occasionally, when the robot sees scraps of paper, it will stop there and utter "Moco!" It will then scan for people nearby while also saying "Moco!"

Fig. 6.4 A new Sociable Trash Box humming "Mocomon"

When someone picks up the trash, it will bend its upper body slightly while saying "Mocomonmon!" The robot will then begin walking again while humming "Mocomon, Mocomon!" (Fig. 6.4).

Thus, the meaning of "Moco!" and "Mocomon!" are not self-contained and are open to interpretation by the surrounding people relative to the circumstances of the specific moment.

For example, "Moco, monmon!" can be interpreted as "thank you," or as words of gratitude. It can also be interpreted as begging, as in "more, more!" or "one more, one more!" People will then pick up the trash while agreeing with their own interpretation. In response to the light bow, people will simply interpret it as a sign of gratitude and feel satisfied at that moment.

Thus, ambiguity is required to allow people to form their own original interpretation to facilitate positive participation and a sense of acceptance, which can also be considered kindness toward the listener.

6.3.2 Rich Communication Created by Incomplete Utterances

Rather than utterances that carry self-contained meaning, incomplete utterances create meaning together by partially opening up the interpretation of the meaning to the listener. This can be considered a relational-oriented action strategy that is common in many weak robots, e.g., Sociable Trash Box or/and iBones robots.

A representative example of the development of the weak robot concept in the communications field includes the Talking-Ally robot, which is characterized by adjusting and organizing the content and timing of utterances using "hearership" that indicate that they are currently listening (Matsushita et al., 2015). In addition, the Talking-Bones robot attempts to tell children old stories; however, it occasionally forgets important words (Onoda et al., 2019).

Conventionally, when communicating with others, there is a need to use only the right amount of fluent spoken language, which indicates the need to properly organize one's thoughts and speak calmly. However, when applying this to communication with a robot, it feels somewhat distant and as though the hearer has been left behind, similar to the utterances generated by a smart speaker system.

The model we used is one in which a child who has just returned from elementary school attempts to tell their mother about the events of their day.

> "Today, I played a lot!" (Oh, with who?)
> "Rei-chan!" (Oh wow, what did you do to play?)
> "I drew!" (Ah, I see.)
> "Cha-chan did too!" (Oh wow, was it fun?)
> "Yup!" ...

Why do children use incomplete utterances in such circumstances? This could be attributed to the children's low language skills. Initially, children cannot organize what they want to say in their heads. They might simply say what they want to convey as they think them.

However, in response to such incomplete utterances, people often ask, "What does that mean?" One strategy is likely to draw out sympathetic engagements (i.e., care) from the surrounding people. It is also thought that the speaker wants to share their happy memories of the day and needs a place where utterances can complement each other.

In the process of investigating such incomplete utterance patterns, we created the Talking-Bones robot, which attempts to tell an old story to children but sometimes forgets important words (Fig. 6.5).

> "A long, long time ago, in a certain place there was a grandpa and grandma." "The grandpa went to the mountains to gather firewood, and the grandmother was at the river..."
> "Uhh, what was it again? What did she go and what did she do...," "Uhh..." ...

It is somewhat strange for a robot to forget important words; however, the child becomes instantly enthusiastic when the robot makes a troubled gesture, e.g., "Hmm?..." or "Uhh, what was it again?"

The child helps out, saying, "didn't she go do the laundry?" In response, the robot states, "Oh, yes! That's it!" "She went to do the laundry."

> "And then the grandma went to do the laundry at the river."
> "Then from the river comes a bobbing and bobbing..."
> "Uh, uhm, what came bobbing down?"
> "Not a watermelon... uhh..." ...

Faced with this unreliable storytelling, the children jump in and attempt to help the robot. By wondering, "what is the robot struggling with" or "what is the robot trying to remember," the child thinks, "it's not this, it's not that." This is essentially putting oneself in the shoes of the other, which is a communication strategy where one attempts to understand the other's situation using one's own thoughts and feelings as clues. The robot and children explore each other's thoughts and feelings about the forgotten word and adjust to each other. In this so-called triadic relationship, there

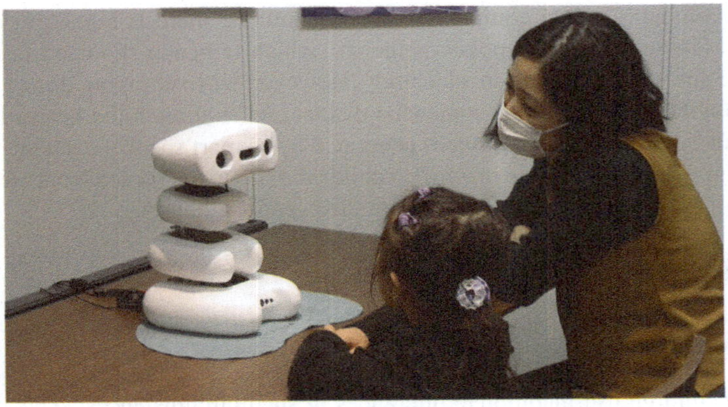

Fig. 6.5 Talking-Bones robot occasionally forgets important words while telling an old story

is a place for being in each other's shoes, where the participants mutually identify themselves with each other.

This triadic relationship can be formed between children and robots and between children alone. The following example presents a disagreement between children in response to the forgetfulness of the Talking-Bones robot relative to the statement, "From the peach . . . uh, what comes out again?" "Ba . . . "

"Isn't it Momotaro?" "Hmm, isn't it a babe?"
"Babe," "N, No . . . ," "Momotaro!"
"Uhmm . . . ," "Babe . . . ," "Oh, baby!"
"Baby! That's it!" (The children laugh.)

The children go back and forth several times over the object that comes out of the peach, and they finally decide to say "Baby!" Thus, here, everyone feels a sense of relief, and laughter occurs.

If the Talking-Bones robot simply read old stories, the children would quickly become bored. Thus, the incompleteness of the robot's utterances (i.e., imperfection in memory and recall) draws out the children's positive involvement and strengths. In addition, even children cannot recite the old story of "Momotaro" to the very end. It can be said that a mutual relationship is formed between the children and Talking-Bones, where each party complements the weaknesses of the others and draws out their strengths.

As a result, compared to when Talking-Bones speaks unilaterally, the communication between the children and the robot would be an enriching activity. This is also expected to lead to a protégé effect, where children ultimately learn things themselves while taking care of a robot.

Here, what is interesting is the satisfied expression on the children's faces when dealing with a clumsy robot. While taking care of robots that require some work, they appear to be somewhat happy. This is similar to the interactions observed between children and our other robots, e.g., the Muu and Sociable Trash Box robots.

It can be said that these will go beyond the conventional values of providing convenience with highly functional robots and realize a state of well-being for the children, i.e., their abilities are fully utilized, and they feel a sense of happiness.

6.4 Conclusion

This chapter has introduced the weak robot concept and discussed ways for humans to interact with social robots.

This concept can also be applied to various simple tools, e.g., scissors. When cutting a thread or string, the hard steel of the scissors compensates for our soft (i.e., weak) human hands. The weakness in our hands requires the help of the scissors and draws out the strength of the scissors.

However, what about the scissors? When simply left on a table, their essential function cannot be realized. It is only in our hands that their function, e.g., cutting a thread, can be exhibited. Thus, the weakness of the scissors is complemented by our freely moving hands. Here, our soft human hands become a strength by using the scissors. The weakness of the scissors changes the weakness of our hands into a strength and draws it out. As a result, tools draw out the strengths of the users while complementing their weaknesses.

Recently, various robots and information devices have been developed in line with the trend of pursuing increasingly convenient objects. As smart entities achieving perfect work, we have sought a high degree of autonomy that allows us to complete work independently. What kind of relationship do we form with such highly convenient systems? Similar to the relationship between an automated driving system and passengers, as soon as we draw a line between a "system that does XX for me" and a "person who gets xx done for me," we become susceptible to falling into a situation of "it just does something without me doing anything!" Here, a disconnect is created between the two entities, empathy for the subject is lost, and we continually escalate our demands, "Quieter! More accurately! More efficiently!"

This pursuit of convenience tends to draw out our arrogance and encourages intolerance. What can we elicit from the previously mentioned well-being perspective? The spread of automated driving systems will be good news for many users, e.g., the elderly and those with disabilities. However, similar to baggage, we will be at the mercy of the decisions of a self-centered automated driving system, including where we are transported. Many of the controls will be entrusted to the system, which therefore does not reach the level of operating freely as desired. In addition, it is difficult to obtain a sense of accomplishment and competence that comes with proficiency, and no sense of unity or connection with the car that has been present to date can be expected. Thus, it appears that we are going backward from a well-being perspective.

We hope that the discussion presented in this chapter will be of some help when considering the symbiosis between advanced robots and humans in the future.

References

Bakhtin, M. (1996). *Слово в романе (translated in Japanese).* HeibonSha.

Calvo, R. A., & Peters, D. (2014). *Positive computing: Technology for Wellbeing and Human Potential.* The MIT Press.

Matsushita, H., Kurata, Y., De Silva, P.R., and Okada, M. (2015). Talking-Ally: What is the future of Robot's utterance generation? Proceedings of the 24th IEEE International Symposium on Robot and Human Interactive Communication (RO-MAN 2015), 291–296.

Okada, M. (2000). Muu: Artificial creatures as an embodied interface, Proceedings of the 27th International Conference on Computer Graphics and Interactive Techniques (SIGGRAPH 2000).

Okada, M. (2012). *Weak Robot (in Japanese).* Igaku-Shoin.

Okada, M. (2017). *Thought of weak robot (in Japanese).* Kodansha.

Okada, M. (2022). *Robot: For symbiotic relationship with humans (in Japanese).* University of Tokyo Press.

Onoda, S., Nishiwaki, Y., and Okada, M. (2019). Interaction design and field study of a forgetful social robot talking-bones, Proceedings of 7th International Conference on Human-Agent Interaction (HAI '19), 259–261.

Saeki, Y. (Ed.). (2017). *Education as caring about the world children care (in Japanese).* Minerva Shobo.

Yamaji, Y., Miyake, T., Yoshiike, Y., De Silva, P. R., & Okada, M. (2011). STB: Child-dependent sociable trash box. *International Journal of Social Robotics, 3*(4), 359–370.

Michio Okada Professor, Interaction and Communication Design Lab. Department of Computer Science and Engineering, Toyohashi University of Technology.

Professor Okada obtained a PhD in Computer Science at Tohoku University. From 1987 to 2005, he was a senior researcher at NTT Basic Research Labs and ATR International. His research interests include human–robot interaction, social robotics, and cognitive science.

Chapter 7
Tourism Engineering for Supporting Stroll—What Is True Travel?

Tomoko Izumi

Abstract Today, the types of tourism are diversifying, and tourists have different needs and their own style of sightseeing. Since tourists can obtain their interesting information about sightseeing area before their sightseeing, they make sightseeing plans to maximize their enjoyment within a limited time schedule before going sightseeing. However, from another point of view, this process seems to lose opportunities for new discoveries and encounters at the local sites of the tourist destination because the tourist follows the pre-planned sightseeing route. That is, it can be said that tourist behavior is restricted by information obtained in advance. In this chapter, the information systems based on Benefit of Inconvenience, which are proposed to generate the lost opportunities for new discoveries, are introduced. However, the great degree of inconvenience would be caused too great an effort to use the system. Therefore, we consider a mechanism with "allowable inconvenience" for sightseeing support. We discuss the allowable inconvenience from various perspectives: in terms of the details of information about sightseeing spots, the locations and timing of presenting information, and the design of presenting textual information about spots.

7.1 What Is True Travel?

I have been working on research to support tourism in the field of information engineering for about 10 years. During this period, my research has focused on tourism navigation systems, tourism information sharing systems, and so on. However, I myself still do not understand that "what tourism is." I face even more difficulty in answering the question, "what is 'true' travel?", which is the subtitle of this chapter.

T. Izumi (✉)
Ritsumeikan University, Kusatsu, Japan
e-mail: izumi-t@fc.ritsumei.ac.jp

H. Kawakami (ed.), *Systems Design Based on the Benefits of Inconvenience*,
Translational Systems Sciences 31, https://doi.org/10.1007/978-981-19-9588-0_7

The history of tourism is said to be as old as the history of human civilization. Tourism styles have changed significantly due to the deregulation of transportation and the development of transportation networks such as railroads and expressway. Furthermore, major changes in tourism styles have occurred over the past 20 years due to the spread of the Internet and the increased use of social networking services (SNS) such as Twitter and Instagram. Today, a huge amount of information about sightseeing spots is available on the Internet, and SNS share information about things and experiences that other tourists have encountered actually in sightseeing spots. Much of such information includes photographs and video clips, which give viewers a visual sense of the sightseeing. In addition, information retrieval and recommendation technologies have also developed, and it is now possible to efficiently obtain information that the viewer prefers from among a huge amount of the information. Because of the availability of these technologies, many tourists often research information on sightseeing spots well and make sightseeing plans to maximize their enjoyment within a limited time schedule before their sightseeing.

Since tourists select sightseeing spots to visit from a vast amount of information, the use of these technologies seems to be enriching tourism. On the other hand, however, the tourist behavior of conforming to a predefined sightseeing plan also seems to lose opportunities for new discoveries and encounters at the local sites of the tourist destination. Before the Internet became widespread, tourists could only obtain information from pamphlets and guidebooks published by sightseeing spots, and they often obtained information on-site and planned their next sightseeing activities. However, advances in information technology have transformed tourism into an act of confirming that the local experience is "surely the same" as the previously obtained information. As a result, opportunities for interaction between tourists and local communities in sightseeing areas, such as communication with the local people and knowing the local environment, may be decreasing. That is, it can be said that tourist behavior is restricted by information obtained in advance.

7.2 Design of Inconvenient Tourist Navigation System

If the opportunities for interaction between tourists and local communities in sightseeing areas are decreased by giving large amounts of information to tourists in advance through the use of information technology, what can be done to regain the opportunities for interaction? The design theory of regaining benefits lost by convenient technologies and services by making them inconvenient is called "Benefits of Inconvenience" (Kawakami, 2019). As an example of system designs based on the Benefits of Inconvenient for supporting tourism, there are system designs that target navigation systems (Takagi et al., 2013a). In these systems, the opportunity for lost interaction is generated by purposely making some of the navigation functions inconvenient. In this section, some of the systems are introduced.

The function that a tourist navigation system should provide is to guide tourists to a desired sightseeing spot. These systems based on Benefits of Inconvenience aim to encourage tourists themselves to become aware of their surroundings through an inconvenient mechanism while providing this functionality. The idea for the inconvenient mechanism is derived from the sightseeing maps installed in sightseeing areas. On sightseeing maps, well-known or landmarked sights are indicated prominently. Thus, the direction and scale vary from map to map, and sometimes even on the same map. That is, the maps have no consistent description. Therefore, tourists look around them and compare the information on the map with their surroundings in order to confirm their own position on the map, the scale and direction of the map. Looking around them means paying attention to what is around them. In other words, this may make it easier for tourists to increase opportunities for interaction with the surroundings and to find interesting spots. The inconvenient tourist navigation systems (Takagi et al., 2013b) are proposed from this ideas.

Nakatani et al. proposed a navigation system using a hand-drawn sightseeing map created by tourists themselves before sightseeing (Nakatani et al., 2011). In the sightseeing using the hand-drawn map, tourists place icons of sights and landmarks they will visit on the electronic map and handwrite their travel routes before sightseeing. When they go sightseeing, the electronic map in the background disappears, and so they move along the route and landmarks that they have drawn. The hand-drawn map, as well as sightseeing maps installed in sightseeing areas, is less accurate than general maps. Thus, the hand-drawn map be expected to have the same effect as sightseeing maps, encouraging tourists to look their surroundings. Tanaka et al. also proposed a sightseeing navigation system that hides a map of a 100-m radius around a tourist (Tanaka & Nakatani, 2010). Because the hidden area is set based on the tourist's current location, the tourist cannot always see the area 100 m around him or her on the map. Of course, the map around the destination is also hidden when the tourist approach it. So, the tourist is required to search for the invisible route or environment on the map. Furthermore, a sightseeing navigation system that does not display maps and does not give any map information at all has been proposed (Takagi et al., 2012a, 2012b). In this navigation system, only the starting point, destination, and landmarks information on the route are provided. Tourists can confirm the appearance of the landmarks by the provided photos, and they can reach the final destination by visiting the landmarks in order. In all of these studies on supporting tourism, detailed map information is intentionally not shown to encourage tourists to take their eyes off the map and look at their surroundings and to promote interaction with the surrounding environment.

However, while these systems provide information to reach a destination, they require significant effort to reach the destination. In fact, in an experiment using a system that did not provide any map information (Takagi et al., 2012b), while it takes about 20 minutes at a typical walking speed to reach a certain destination by the shortest route, it took some collaborators two hours to reach that destination. While free exploration in sightseeing areas without the constraints of time is appealing, these cases were taking too long time. Hiraoka et al. provide a definition of Benefits of Inconvenience in (Hiraoka & Kawakami, 2019), in which they state that Benefits of Inconvenience must have a subjective, objective effort and a subjective, objective benefit. The systems introduced in this section satisfy this definition of Benefits of Inconvenience, because, while they require more physical and psychological effort to reach the destination, they also have the benefit of enabling tourists to encounter various sights and experience enjoyment. However, if the effort required before the benefits are realized as great, it is difficult to motivate people to use these systems.

7.3 Sightseeing Support using Allowable Inconvenience

From the studies introduced in the previous section, it is shown that the introduction of an inconvenient mechanism into a tourist navigation system increases the interaction between tourists and surrounding environments and encouraging discoveries. However, if the effort for tourists caused by the inconvenient mechanism increased significantly, it would be difficult to motivate them to use the mechanism. On the other hand, if a convenient system is used to efficiently provide tourists with sufficient information, their behavior will be limited to the activities suggested by the system. Because of this observation, our research group has been considering a mechanism with "allowable inconvenience" that increases opportunities for interaction between tourists and the surrounding environment while providing a certain degree of convenience through system functions. This section discusses the allowable inconvenience from various perspectives: in terms of the details of information about sightseeing spots, the locations and timing of presenting information, and the design of presenting textual information about spots.

7.3.1 Allowable Inconvenience for Information of Spots

In this section, we consider a system that recommends surrounding sightseeing spots to tourists who enjoy strolling. Even in such a system, giving tourists detailed information about recommended spots will encourage them to visit the recommended spots, which seems that their behaviors is limited by the information. In contrast, if tourists are given too little information, since the effort to find interesting spots is high, various behaviors of tourists cannot be induced and new interactions cannot be encouraged.

From this motivation, we conducted a comparative study to determine the level of detailed information that would encourage tourists to change their behavior, but not take limited actions such as only visiting specific recommended spots (Hiraishi et al., 2018). For this purpose, we divided the information on sightseeing spots into two categories, that is, characteristics information and location information, and defined four levels of information detail for each categories. The characteristic information on recommended spots is classified based on the information written in general guidebooks to introduce sightseeing spots. Specifically, the information is

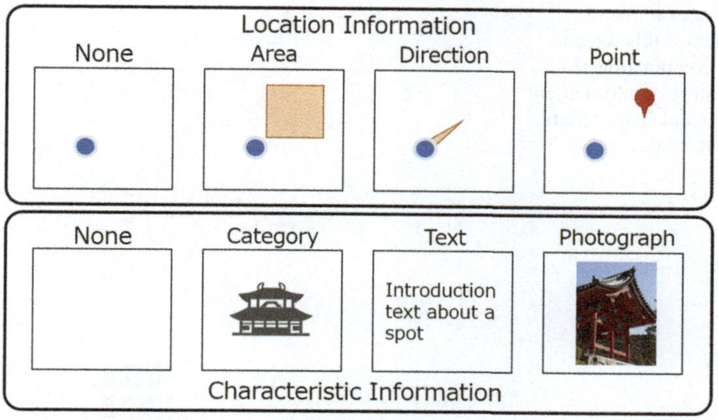

Fig. 7.1 The categories of the characteristic and location information.

classified by photographs or text and categories that indicate what type the sightseeing spot is. Photographs give tourists a concrete visual image of a sightseeing spot and thus have the highest level of detail in information. Text is the second most detailed because, although tourists have difficulties understanding the visual image of a spot from the text, they can guess about the spot from the text. The category shows the type of a spot, such as temples, shrines, cafes, etc., and so has the lowest level of detail. For location information, the levels of detail is set to the degree that location of a spot can be limited. That is, we set three types of information, a point for the exact location of the spot, a direction, and a range of the area in which the spot is located. For each category, we classified the information into four levels of details, including the case in which no information is given to a tourist. Figure 7.1 shows these categories of the characteristic and location information.

The comparative verification was conducted on 16 patterns of information presentation combining characteristic and location information. We asked the participants of the experiment to use each system of the pattern for sightseeing and analyzed their behavior. In the experiment, recommendation spots were selected by a recommendation method based on collaborative filtering using NMF (Non-negative Matrix Factorization). Twenty-four university students (18 males and 6 females) participated in the experiment. To enjoy sightseeing through conversation with a partner, we paired the participants. A total of 12 pairs were asked to explore a sightseeing area in Kyoto, Japan using one of the systems. Figure 7.2 shows an example of system screens in which it provides area and category information of a recommended spot. The location information is displayed in the upper half of the screen and the characteristic information in the lower half. We evaluated the impact of the system on the behavior of the participants based on their travel routes, observation of their behavior, and the results of their responses to the questionnaire. The results showed that a pattern combining area information indicating the approximate location of a spot and category information indicating

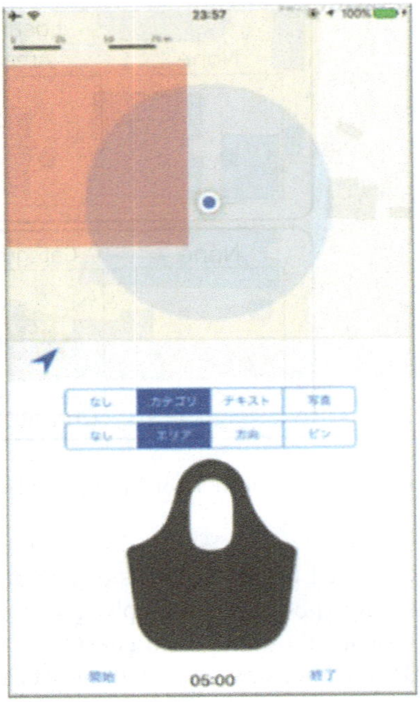

Fig. 7.2 An example of system screens: this screen shows the combination of area and category information for location and characteristic information, respectively.

the type of spot may guide the participants not only to the recommended spots but also to various other spots.

7.3.2 Allowable Inconvenience for Locations of Providing Information

As mentioned above, sufficient information about sightseeing spots is available on SNS and tourist information websites. Thus, before visiting a sightseeing area, most tourists research the sightseeing spots they are interested in and select spots to visit. In this case, tourist behavior is limited within a tourism plan because tourists follow the well-developed plan in advance. If tourists are not given information on sightseeing spots, they do not know what spots are available where. The tourist's behavior will be limited only to the area they coincidentally visit, and in this case, their behavior is also limited. Hence, we proposed to restrict locations to share information about sightseeing so that tourists are given information on tourist spots but not restricted in their activities (Takagi et al., 2013b).

Since tourists are restricted in their behavior by obtaining information before sightseeing, we consider making it impossible for them to obtain information before sightseeing. In other words, the tourists obtain information about sightseeing spots on-site at sightseeing area. There are various means through which information

Fig. 7.3 A journey note placed in a sightseeing spot.

is provided by sightseeing area, such as pamphlets, billboards, sightseeing maps, etc. The difference between these means and SNS is whether or not information is shared with other tourists. Information about actual experiences of other tourists is very useful for tourists who are deciding their future sightseeing activities. One existing tool for sharing information among tourists on-site at sightseeing spots is the "Journey note" (Fig. 7.3). This is simply a notebook placed in hotels and other facilities at sightseeing spots, and any tourist who visits the spot can freely write in and read the notebook. The minor information in the notebooks is interesting to the tourists who visit the spot, and unique communication is taking on among the tourists who read and write in the notebook. By restricting the locations where information is available, the value of the information is increased, and tourists are more interested in the information.

The proposal is described in detail. A virtual journey note is set up at a certain location, and access to that note is restricted to only those within a radius of a few tens meters from that location. There are no constraints on the location of the journey notes, but we assume that they will be placed at stations, locations of sightseeing maps, stores, and so on. Tourists who are within the accessible area can submit information to the virtual note and can view the shared information using their smartphones and other devices. They can also reply to shared information with reactions and comments. Thus, by sharing information locally at a sightseeing spot, it encourages tourists to stroll around the sightseeing area to obtain information, and then to stroll further based on the information they obtain there.

7.3.3 *Allowable Inconvenience for Timing of Providing Information*

The journey note introduced in the previous section only considers restrictions on the locations of providing information and does not consider the flow of the sightseeing. Tourists often set their main destinations, which are mostly well-known sightseeing spots, for sightseeing. They use transportation to get near their destinations, walk around the area, go to the destinations, and when they finish their visit, they move on to the next destination. Many tourists repeat this flow. In this section, we consider the location and the timing of providing information considering this flow of the sightseeing. Concretely, we proposed to delay the timing of providing information until after reaching of the main destination (Izumi & Takemoto, 2019).

Figure 7.4 indicates the flow of sightseeing when the providing information of sightseeing spots is delayed until after the visit to a destination. In this section, we consider a tourist strolling the area surrounding a destination spot. In many sightseeing areas, there are many souvenir stores and restaurants surrounding well-known sightseeing spots. Then, tourists visit the famous spot as their destination and explore the surrounding area. For such sightseeing area, the journey note in the previous section is set at the destination sightseeing spot. That is, tourists are given the restriction of being able to share and view information about surrounding spots only at that destination spot. In other words, they can obtain information on sightseeing spots in the surrounding area only after arriving at the destination. It can be expected that, after arriving at the destination, information about interesting spots that have not yet been visited will encourage additional sightseeing behavior. Furthermore, the information shared at the destination spot is constrained to be collected within a certain distance to the spot. That is, when tourists enter within a certain distance from the destination, they can collect information to be shared at that destination. In order to collect information before visiting a destination, which is a place for information sharing, tourists stroll around with no prior knowledge at the time of information collection. This approach is believed to encourage tourists to stroll the surrounding area to share valuable information at

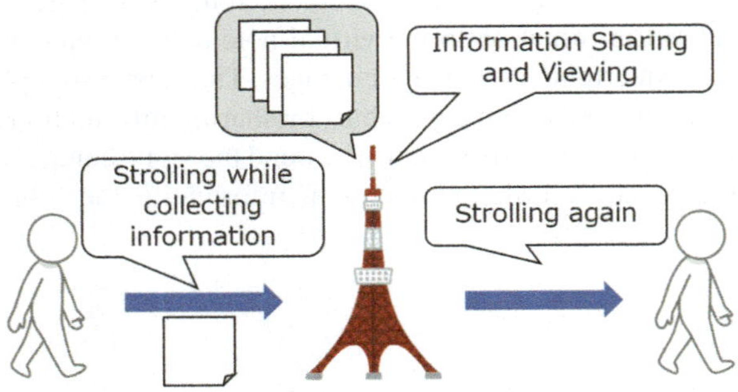

Fig. 7.4 Mechanisms for delaying the timing of information presentation.

their destination sightseeing spots, because of the desire for approval seen in SNS. Furthermore, obtaining new information at the destination could encourage strolling behavior again on the return way from the destination. Experiments were conducted around Kiyomizu-dera Temple and Gion area in Higashiyama-ku, Kyoto, Japan, using a prototype system that realizes this proposal. The results indicate that some participants made decisions about their next sightseeing activities based on the shared information at their destinations.

7.3.4 Allowable Inconvenience for Presenting Textual Information about Spots

One of the on-site information providers for tourists is an information board that introduces sightseeing spots. These information boards introduce the historical background and characteristics of the spots. Information boards can provide information to tourists on-site at the spots and are easy to introduce. However, the information boards installed in temples, shrines, and other historical buildings in Japan are very difficult for tourists to read because many of the words on the boards are unfamiliar to them. For this reason, some tourists have low motivation to read the text on the information boards.

In order to get tourists interested in reading the text on the board, a method of providing information that would attract the interest of them is necessary. For this purpose, it is better to provide information in a way that includes interaction with tourists and to present content that will attract their interest. One way to achieve this is to change the content on the board, but in our study (Izumi et al., 2020), we discussed a method to change only the presentation method without changing the content of the text on the original board.

We consider an interactive system in which words derived from the original text presented sequentially. More concretely, the next word is output when a visitor taps on the screen of a touch-panel digital signage. However, presenting the sentences in the order in which they are originally written is not enough to attract the interest of tourists. The proposed system presents words of interest to tourists as keywords first, in order to attract their attention even if the sentences contain words that are difficult for them to understand.

In the following, please note that this study is for the Japanese language. There are researches on interest in words for catchphrases or information exchanged in online chats. Catchphrases are similar to the role of the words we focus on in this study, because they need to attract the reader's interest in a short sentence. The results of the linguistic engineering analysis conducted by Yamane and Hagiwara (2012) show that catchphrases often have part-of-speech sequences such as "noun-particle-noun" and "noun-particle-verb." Ishii et al. (2004) discuss topics of interest to users in the message sequence for online chat. This study focuses on noun phrases connected by "no" (in Japanese, which is similar to the word "of'" in English) between nouns and proper nouns as words that attract the user's interest. In these studies, nouns and noun-phrase words are used to attract viewers' interest. Thus, as a preliminary experiment, we extracted nouns and noun phrases containing proper nouns from temple and shrine information boards to investigate which words would attract the interest of tourists. The results indicate that noun phrases, which are combinations of verbs, adjectives, and nouns and nouns, are more likely to attract tourists' interest than proper nouns.

Based on the results of the preliminary experiment, our proposal is to present the noun phrases in an information board as keywords at the beginning. However, a tourist cannot understand what the original sentences express through the noun phrases only. Then, when the tourist selects the noun phrase presented first, other words related to the phrase are presented to help the tourist understand the meaning of the original sentence. In the related studies (Yamane & Hagiwara, 2012; Ishii et al., 2004), nouns are also shown as words of interest to users. Then, in our proposal, after one of the noun phrases is selected, other nouns in the same sentence are displayed. Furthermore, when that noun is selected, the verbs in the same sentence are displayed.

It would be difficult to understand the content of the text if the words in the text are presented in the order that tourists' interest, as in the proposed method. On the other hand, when words are presented in the order of their appearance in the original text, a tourist could easily understand the structure of the text, but the tourist would not be very attracted to it. We therefore conducted a comparative experiment using three systems to verify the degree of readability and attractiveness. The first system separates the sentences into phrases and provides the words one by one in the order they appear in the text on the information board. In the first system, the tourists easily understand the structure of the sentence because the words on the screen are always single and are provided in the order of appearance in the original board. The second system divides the text on the board into sentences, delimits each sentence into phrases. Then, for each sentence, the system presents the separated words in the order of their appearance in the original sentence. In this case, the words are presented for each sentence, so there are as many words on the screen as the number of sentences. In the system, the structure of individual sentences is easy to understand, but the structure of the whole texts is difficult to understand. The third system is based on the proposed method: Words of noun phrases of interest to the tourists are presented first, and then words related to those words are subsequently presented (Fig. 7.5). In this system, the order in which words appear is different from the order in which they appear in the original text. So, it is difficult to understand the content of the original text. However, since a large number of interesting words are displayed at the beginning, tourists may actively try to read and understand the content.

We conducted a comparative experiment using these three systems (Izumi et al., 2020). The results showed that the second system, in which words are displayed for each sentence, may make the user feel enjoyable and understandable of the sentences. The third system was evaluated as the most difficult to understand the contents of the original texts. At the same time, however, the third system showed

Fig. 7.5 Signs presenting only keywords of interest. The characters in the circles are keywords from the text written in Japanese

the possibility of encouraging the tourists to check many words and to take action to find interesting words among them.

7.4 Conclusions

This chapter introduces a tourism support system that includes an inconvenient mechanism. The existing studies introduced inconvenient mechanisms to increase opportunities for interaction between tourists and the surrounding environment. However, there were problems with the large amount of effort required of tourists caused by this mechanism which prevented them from being motivated to use the system. In contrast, this chapter presented case studies of how to support tourism through allowable inconvenient mechanisms that also provide some convenient functions.

Today, the types of tourism are diversifying, and tourists have different needs and "true travel" in terms of what they are looking for in tourism. The development of information systems provides tourists with high-performance functions, but the wrong direction or degree of support may interfere with the tourists' original trip. I believe that supporting tourists without restricting their actions or removing their choices of action is supporting their "true travel." In the field of behavioral economics, a mechanism for voluntarily guiding people in a desirable direction, rather than forcing them to do so, is called a nudge (Oliver & Ubel, 2014; Yamane, 2014). Similarly in tourism support, a system in which tourists voluntarily work on the surrounding environment to find new discoveries and opportunities is desirable, because it is an approach to support each tourist without changing their own travel style. I suggest that one of these approaches may be allowable inconvenient mechanism. While many tourism support systems and applications have been proposed, how to realize such an allowable inconvenient mechanism is still an open question.

References

Hiraishi, Y., Kitamura, T., Izumi, T., & Nakatani, Y. (2015). Experimental verification of sightseeing information as a weak trigger to affect tourist behavior. In *Proceedings on International Conference on Social Computing and Social Media, the 20th International Conference on Human-Computer Interaction (HCI' 18)* (pp. 303–317). Springer.

Hiraoka, T., & Kawakami, H. (2019). Redefinition of benefits of inconvenience. In *Proceedings on the 21th International Conference on Human-Computer Interaction (HCI' 19)* (pp. 131–144). Springer.

Ishii, M., Nakawatase, H., & Tomita, J. (2004). Topic extraction from a message stream using noun phrase and word pressure around the latest message. *IPSJ SIG Technical Report (2004-NL-160), 2004*(23), 79–84 (in Japanese).

Izumi, T., & Takemoto, K. (2019). Proposal for encouraging tourists to stroll by restricting information sharing location to destinations. In *Proceedings on the 21th International Conference on Human-Computer Interaction (HCI' 19)* (pp. 295–306). Springer.

Izumi, T., Guan, X., Kitamura, T., & Nakatani, Y. (2020). Proposal for information presentation to attract tourists' interest in an installation-type tourist information provision system. In *Proceedings on Annual conference of SICE SANDI*. SICE (in Japanese).

Kawakami, H. (2019). Designing human-machine systems focusing on benefits of inconvenience. In A.-H. Sato (Ed.) *Applications of data-centric science to social design* (pp. 51–62). Springer.

Nakatani, Y., Tanaka, K., & Ichikawa, K. (2011). New approach to a tourist navigation system that promotes interaction with environment. In M. Schmidt (Ed.), *Advances in computer science and engineering* (pp. 353–370). IntechOpen.

Oliver, A., & Ubel, P. (2014). Nudging the obese: A UK-US consideration. *Health Econ Policy Law, 9*(3), 329–342.

Takagi, S., Izumi, T., & Nakatani, Y. (2012a) A sightseeing navigation system without route information. In *Proceedings on International Conference on Advances in Human-oriented and Personalized Mechanisms, Technologies, and Services* (pp. 50–55). IARIA.

Takagi, S., Masuda, M., Izumi, T., & Nakatani, Y. (2012b). The tour navigation system using the landmarks that are customized by personal preferences. In *Proceedings on International Symposium on Socially and Technically Symbiotic Systems* (pp. 47-1–47-7).

Takagi, S., Izumi, T., & Nakatani, Y. (2013a). Five sightseeing navigation systems using limited information. In *Proceedings on 2013 Conference on Technologies and Applications of Artificial Intelligence (TAAI' 13)* (pp. 312–317). IEEE.

Takagi. S., Izumi, T., & Nakatani, Y. (2013b). Sightseeing assistance via information inheritance with journey note system in sightseeing spots. In *Proceedings on The World Congress on Engineering and Computer Science (WCECS 2013)* (pp. 764–769). IAENG.

Tanaka, K., & Nakatani, Y. (2010) Sightseeing navigation system that promotes interaction with environment by restricting information. In *Proceedings on 2010 IEEE International Conference on Systems, Man, and Cybernetics (SMC 2010)* (pp. 1407–1413). IEEE.

Yamane, S. (2014). Shikake as a nudge. *Journal of Artificial Intelligence Society, 28*(4), 596–600 (2014).

Yamane, H., & Hagiwara, M. (2012). Feature analyses for catchphrases using multiple corpora. *Journal of Japan Society of Kansei Engineering, 11*(2), 233–239 (in Japanese).

Tomoko Izumi received the B.E., M.E. and D.I degrees in computer science from Osaka University in 2003, 2005, and 2008. She is an associate professor of College of Information Science and Engineering, Ritsumeikan University. Her research interests include design of human-computer interaction for supporting sightseeing, or evacuation in a disaster situation, recalling memories. She is a member of IEEE, HIS, IEICE, and IPSJ.

Chapter 8
Application of Fuben-eki to Travel Commerce and its Large-Scale Social Implementation

Tomohiro Shirakawa

Abstract Fuben-eki refers to the benefits obtained by daring to accept inconvenience, and Fuben-eki System Theory is a design theory of products, services, and systems based on Fuben-eki. Fuben-eki System Theory actively acknowledges the value and significance of inconvenience and is the antithesis of the uncritical pursuit of rationality and elimination of waste. In this respect, the Fuben-eki System Theory is compatible with travel, especially with a type of travel that does not have a specific goal in mind. In this study, we rethought the significance of travel from the perspective of the Fuben-eki System Theory in order to transform the meaning of travel itself and create new value in travel. As a result, a new travel commerce/social shopping application, HAKOBIYA, was devised. HAKOBIYA successfully accomplished a number of downloads and a very large gross merchandise value, and it became the first large-scale social implementation of a business model based on the Fuben-eki Theory.

Keywords Fuben-eki · Travel · Travel commerce · Social shopping · HAKOBIYA

8.1 Introduction

Fuben-eki refers to the benefits obtained by daring to accept inconvenience, and Fuben-eki System Theory is a design theory of products, services, and systems based on Fuben-eki (Kawakami, 2009, 2011, 2017; Kawakami et al., 2017). Fuben-eki System Theory actively acknowledges the value and significance of inconvenience and is the antithesis of the uncritical pursuit of rationality and elimination of waste. In this respect, the Fuben-eki System Theory is compatible with travel, especially with a type of travel that does not have a specific goal in

T. Shirakawa (✉)
Department of Information and Management Systems Engineering, Graduate School of Engineering, Nagaoka University of Technology, Nagaoka, Japan
e-mail: shirakawa@vos.nagaokaut.ac.jp

H. Kawakami (ed.), *Systems Design Based on the Benefits of Inconvenience*,
Translational Systems Sciences 31, https://doi.org/10.1007/978-981-19-9588-0_8

mind. This is because chance encounters, such as chance discoveries triggered by side trips, which occur because the journey is not optimized, are truly Fuben-eki. In fact, there have already been several products and services that attempt to provide opportunities for new experiences by applying the Fuben-eki System Theory, such as the following:

- Blurring Navigation (In Japanese, "Kasureru Navi") (Kawakami et al., 2017; Kitagawa et al., 2010; Kawakami, 2012).

  Unlike normal navigation applications that display the user's current location along with a map, the map becomes blurred and obscured for places the user has passed several times, gradually losing information on the map that gives a third-person, bird's-eye view. This inconvenience makes the first-person experience an active one, and the user benefits from being able to memorize streets and roads more easily.
- Left Turn Only Tour (Anon, n.d.)

  As the name implies, the tour prohibits right turns and only going straight or left turns are allowed while traveling. Since right turns are prohibited, the tour is inconvenienced by the fact that left turns must be repeated three times in order to achieve a route that requires a right turn. This is beneficial in that it provides an opportunity to pass through extra detours that would not be taken in a normal route selection, and offers encounters and discoveries that would not be available in a normal situation. However, the fact that three left turns can be substituted for a right turn is based on Kyoto's grid-like structure.

Many other products and services based on similar ideas have been proposed (Japan Industrial Designers Association, 2019). All of these products and services introduce some kind of inconvenience or trouble to travel, or conceal some of the information related to travel, thereby promoting chance encounters during travel and transforming the interaction between the traveler (user) and the objects that appear in the travel. However, to summarize, all of the above are for "daring to enjoy inconvenience," and have not led to the transformation of the meaning of travel itself or the creation of new value in travel. Therefore, existing ideas may be viable as small-scale products or services, but they cannot become new business models.

In this study, we thus rethought the significance of travel from the perspective of the Fuben-eki System Theory in order to transform the meaning of travel itself and create new value in travel. As a result, a new travel commerce/social shopping application, HAKOBIYA, was devised, and it became the first large-scale social implementation of a business model based on the Fuben-eki Theory.

8.2 Development of HAKOBIYA

8.2.1 The Inspiration for HAKOBIYA: Monetization of Fuben-eki

As introduced in the previous chapter, Fuben-eki has already been realized by introducing inconvenience to travel and transforming the user experience in the form of daring to enjoy the inconvenience. However, these services do not create a new niche for travel, and as already mentioned, they may become small-scale products or services, but they do not lead to the creation or proposal of new business models. What kind of inconvenience, then, would enable the transformation of travel itself, the creation of new value in travel, and even the creation of a new business model?

Mobility and the accompanying encounters with the unknown are the main components of travel. In fact, the strategy of promoting chance encounters by making travel inconvenient has been a common strategy in previous studies on the Fuben-eki of travel, and this point will be followed in this study. In addition, this study begins by monetizing the inconvenience of travel. This is the core idea of this study, and it is a method that has rarely been assumed for conventional Fuben-eki methods.

Although there have been several attempts to define the exact definition of Fuben-eki (Kawakami et al., 2017; Nishimoto, 2019; Hiraoka & Kawakami, 2019), all of them basically argue for a form in which an (unexpected) benefit is generated in a secondary task by undertaking an inconvenience in a primary task. On the other hand, the idea of monetizing the inconvenience of travel gives a new main objective of acquiring money, whereas travel was originally the main objective (and in fact, which is the main objective depends on the attitude of the user). This is probably the reason why the idea of monetization of Fuben-eki did not arise from the standpoint of the Fuben-eki System Theory. In fact, depending on which proposal for the definition of inconvenience is adopted, the monetization of inconvenience by HAKOBIYA may fall outside the definition of Fuben-eki. However, as will be discussed later, it is certain that HAKOBIYA causes Fuben-eki phenomena in various aspects.

How is it possible to monetize inconvenience? For example, if we add restrictions to travel, such as the requirement to meet a certain person at a certain place at a certain time and date, where is the potential for financial rewards? This becomes possible by having individuals take charge of international logistics. The details of this mechanism are described in the next section.

8.2.2 How HAKOBIYA Monetizes Inconvenience

HAKOBIYA is a social networking service for travel commerce and social shopping (Fig. 8.1). The traveler transports the goods purchased mainly in his/her home

Fig. 8.1 The home screen of HAKOBIYA

country to the client during the trip. Communication between the client and the traveler during the trip is made possible by HAKOBIYA's chat function, and the payment of remuneration is also made possible by the functions of the application.

The HAKOBIYA traveler accepts the travel restriction that he or she must include in the trip a schedule of meetings with a specific person at a specific place and at a specific time. In return, the client pays the traveler the market price of the desired item plus an additional 20–40% (the percentage varies depending on the urgency and intensity of the request). In many cases, even with this additional percentage, the cost to the client is less than the cost of purchasing goods via international transportation, and the product can be obtained more quickly. In this way, the inconvenience of travel restrictions is monetized (Fig. 8.2).

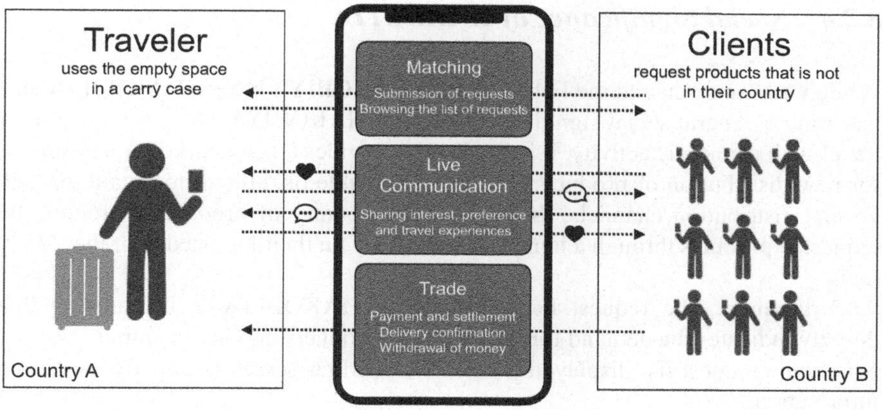

Fig. 8.2 The conceptual design of HAKOBIYA

8.2.3 Fuben-eki Associates with HAKOBIYA

HAKOBIYA provides the direct benefit of monetary rewards for accepting the inconvenience of travel routes and itineraries, but it also provides several other associating Fuben-eki.

The first such Fuben-eki is the benefit of chance encounters associated with the inconvenience of travel. The clients who purchase goods do not necessarily reside in major cities, nor do they reside in the vicinity of landmarks that are the possible object of sightseeing. Therefore, the benefit is that it promotes encounters in the sense that it provides opportunities and possibilities to visit places that would not normally be considered as stopover destinations in a normal sightseeing tour.

Second, and this is a benefit to both the traveler and the client, HAKOBIYA encourages communication between the traveler and the client, because the transaction is based on a personal relationship between the traveler and the client. It also provides a benefit because the communication is based on an international transaction. The traveler has an opportunity to learn about the demand for his/her products in other countries, and to deepen his/her understanding of his/her own country or place of residence from another country's point of view. By sharing the process of acquiring the product with the traveler, the client can not only purchase the product but also learn the context and cultural background of the product, such as its position in the country of origin and the degree of rarity of the product.

Third, because the traveler travels while communicating with the client, i.e., the local people at the destination, the traveler can obtain advice on the itinerary and tourist information during the trip. This is a valuable source of firsthand local information about the destination.

8.2.4 Social Significance of HAKOBIYA

While we have discussed the Fuben-eki that HAKOBIYA brings, HAKOBIYA also has a more general social significance: because HAKOBIYA offers a new form of travel with economic activity, it immediately provides a new market. It also allows for new distribution of products that are unavailable or difficult to obtain through normal distribution channels, as locals in the country of production procure the requested products through a human search based on their knowledge of their daily lives.

Furthermore, the request list presented in HAKOBIYA is information that directly indicates the demand for the country's products in other countries. In other words, the request list displayed on the HAKOBIYA screen is directly marketing information.

Finally, HAKOBIYA is a social networking service and does not mediate only one-off transactions. Therefore, there may be cases in which a "client" relationship is established in which a traveler repeatedly makes a request to the same person. In addition, the application will be upgraded in the future to enable transactions in which the traveler not only responds to the client's request but also makes his/her itinerary public first and the client makes a request according to the itinerary. In this way, influencers who have a strong influence on other users (and who are supported by many clients) can be expected to appear, just as in the case of ordinary social networking services. As these exchanges are repeated and continued, clusters with shared tastes and values will be formed. In other words, HAKOBIYA has a great potential to create a new global community in the future.

As described above, HAKOBIYA is a service that not only realizes the Fuben-eki of travel from multiple perspectives but also has more general social significance.

8.3 Current Status and Accomplishments to Date of HAKOBIYA

This chapter presents the results obtained from the operation of HAKOBIYA, based on currently available statistical data. The data shown are slightly out of date, but this is because data for FY2021 and beyond are not publicly available. Note that the authors were involved only in the conceptual design of HAKOBIYA, and that the development, release, and operation of the application were conducted by Fore-Co., Ltd. (Founder CEO: Yuto Tanaka, Co-founder COO: Kenshiro Nakagawa).

Figure 8.3a shows data on the number of HAKOBIYA downloads. Since its release in July 2019, the number of HAKOBIYA downloads has been steadily increasing, and as of June 2020, the total number of downloads worldwide is approximately 150,000. Figure 8.3b shows the total number of HAKOBIYA transactions from January 2020 to June 2020. The total value of transactions in circulation has generally increased in tandem with the increase in the number of

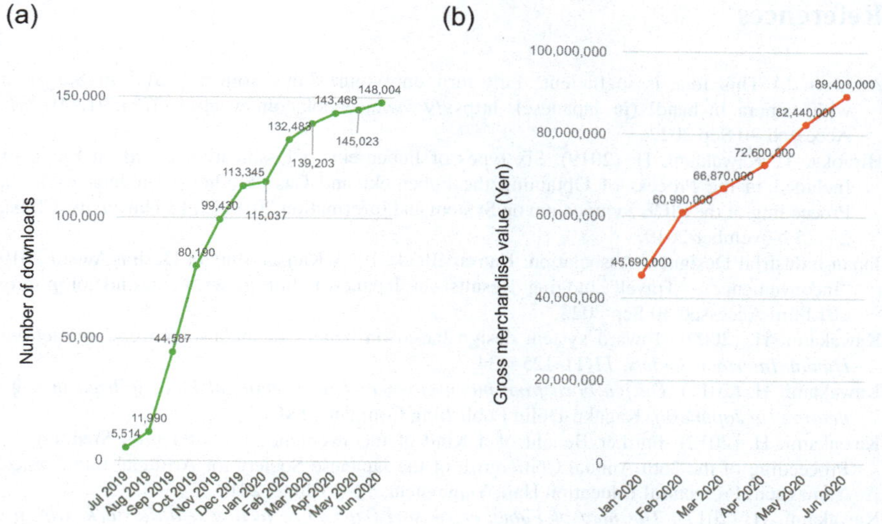

Fig. 8.3 (**a**) Number of HAKOBIYA downloads (from January 2019 to June 2020) (**b**) Gross merchandise value of HAKOBIYA (from January 2020 to June 2020)

downloads, and the travel commerce market created by HAKOBIYA is expected to grow to over 1 billion yen in the future. As indicated by the above data, HAKOBIYA is already widely used, and its scale is still growing. In other words, HAKOBIYA is the first large-scale social implementation of a business model based on the Fuben-eki System Theory.

8.4 Conclusion

In this study, HAKOBIYA, a travel commerce and social shopping application, was developed based on the idea of exploring new ways of travel and new ways to benefit from the inconvenience of travel by monetizing the Fuben-eki. The concept design of HAKOBIYA, which was implemented by the author, was socially implemented by Fore-Co., Ltd. and has been accepted worldwide, establishing a relatively large market. HAKOBIYA is the first large-scale social implementation of a business model based on the Fuben-eki System Theory.

References

Anon (n.d.) This tour is inefficient! Left turn only tour! Let's collect "lefts" in Sakyo-ku with camera in hand! (in Japanese). https://www.facebook.com/events/1607345912902837/ Accessed 30 Sep 2022.

Hiraoka, T., Kawakami, H. (2019). Six types of Fuben-eki—Classification based on Elements Included in the Process of Obtaining the Fuben-eki and Case Studies - (in Japanese). In: Proceeding of the SICE Symposium on System and Information 2019, Chiba University, Chiba, 23–25 November 2019.

Japan Industrial Designers Association, Kansai Block: JIDA Kansai Student Design Award 2019 "Inconvenience x Travel" Judging Results (in Japanese). http://www.kansai.jida.or.jp/info/101.html Accessed 30 Sep 2022.

Kawakami, H. (2009). Toward system design based on benefit of inconvenience. *Transactions Human Interface Society, 11*(1), 125–134.

Kawakami, H. (2011). *Design born from inconvenience: Unconventional ideas utilized in engineering (in Japanese)*. Kagaku-Dojin Publishing Company, INC.

Kawakami, H. (2012). Further Benefit of a Kind of Inconvenience for Influential Medium. In: Proceeding of the 26th Annual Conference of the Japanese Society for Artificial Intelligence, Yamaguchi Prefectural Education Hall, Yamaguchi, 12–15 June, 2012.

Kawakami, H. (2017). *The idea of Fuben-eki ~ sorry, if you're feeling a little stuck, why not embrace inconvenience? (in Japanese)*. Impress Corporation.

Kawakami, H., Hiraoka, T., Okita, M., Taniguchi, T., Shiose, T., Okada, M., Izumi, T., Nakatani, Y., Nishimoto, K., Suto, H., & Shirakawa, T. (2017). *Fuben-eki—Design of systems that take time and effort (in Japanese)*. Kindai Kagaku-Sha.

Kitagawa, H., Kawakami, H., Katai, O. (2010). Degrading navigation system as an explanatory example of "benefits of inconvenience" (in Japanese). In: Proceedings of SICE Annual Conference 2010, Grand Hotel, Taipei, 18–21 August, 2010.

Nishimoto, K. (2019). A Personal Proposal on the Definition of Fuben-eki and Design Guidelines for Fuben-eki Systems (in Japanese). In: Proceeding of the SICE Symposium on System and Information 2019, Chiba University, Chiba, 23–25 November 2019.

Tomohiro Shirakawa received his B.S. and M.S. degrees from Osaka University, Japan, in 2002 and 2004, respectively, and Ph.D. degree from Kobe University, Japan, in 2007. From 2007, he worked for Tokyo Institute of Technology for three years as a postdoctoral fellow of the Japan Society for the Promotion of Science. From 2010 to 2021, he was with National Defense Academy of Japan and served as a research associate and a lecturer. Since 2021, he has been an associate professor of the Department of Information and Management Systems Engineering, Graduate School of Engineering, Nagaoka University of Technology, Nagaoka, Japan. He is engaged in interdisciplinary research between life/human science and intelligent informatics.

Chapter 9
Product Design in "Post-Smart Era": Explanation of the Student Competition of Japan Industrial Design Association as an Example

Tomoaki Kageyama

Abstract Under the banner of "Smartification," the industrial world has been developing products to create a society and lifestyle that is reduced time and labor. On the other hand, the concept of "benefit of inconvenience" which focuses on the positive aspects of time and labor experiences that are at odds with smartification, is attracting attention. Furthermore, there is a movement in Japan that favors a lifestyle that takes time and labor, known as "Teineina-kurashi (in Japanese)." The author named the recent trend of pervasive values that are at odds with smartification "Post-Smart," based on the Post-modernism that occurred as a reaction to Modernism in the 1970s. In this study, I focused on the benefit of inconvenience as a necessary design concept in the post-smart era, and I organized requirements for creating the benefit of inconvenience. Then, I conducted design work for university students majoring in product design to create products with the benefit of inconvenience. As a result, we could create several product ideas that won prizes in a design competition on the theme benefit of inconvenience.

Keywords Product design · Essence of designing · Post-smart · Benefit of inconvenience · Tolerance for inconvenience

9.1 Essence of Design

Nowadays, the word "design" has permeated the world and can be seen and heard in various situations. On the other hand, there is still a misconception that "design is merely the act of beautifying the appearance of an object." Product design, as the name implies, is the design of industrial products. In the first part of this paper,

T. Kageyama (✉)
Graduate School of Design and Architecture, Nagoya City University, Nagoya-shi, Aichi, Japan
e-mail: t_kage@sda.nagoya-cu.ac.jp

H. Kawakami (ed.), *Systems Design Based on the Benefits of Inconvenience*, Translational Systems Sciences 31, https://doi.org/10.1007/978-981-19-9588-0_9

I explain the essence of product design to establish a base for "utilization of the benefit of inconvenience" in the practice of product design.

9.1.1 An Example of the Good Design Award in Japan

When discussing the essence of product design, a product often picked up is the "NANOPASS 33" insulin needle, which won the Grand Prize at the 2005 Good Design Awards in Japan. This syringe has no significant novelty in terms of appearance. However, the company created a "painless" syringe by devising an ingenious needle shape. This value of painlessness is a value for human beings in minds, and the process developed to make such value has been recognized as a good design (Japan Institute of Design Promotion, 2005).In other words, the essence of product design is to create new values for human beings and society (Japan Institute of Design Promotion, n.d.). As the product design adage "Form Follows Function" suggests, beautiful form is a secondary value that emerges in the pursuing value for human beings and society, not an essential value.

9.1.2 Exploration of Essential Value

As I mentioned earlier, the essence of design is to create new value for human beings and society, so it is vital to find the essential value in the age. This section considers the negative aspects of convenience, referred to as the "Harm of convenience" in a study of the benefit of inconvenience.

The upper part of Fig. 9.1 is a conceptual diagram showing the evolution of remote communication tools in Japan, with the vertical axis representing the ratio of each media (Ministry of Internal Affairs and Communications, n.d.; Hakuhodo Institute of Life and Living, n.d.). From the days when letters, telephone calls, and faxes were the mainstream, the main tools of remote communication shifted to e-mail when NTT DoCoMo launched the cell phone e-mail service in 1997. In addition, starting with "mixi" in 2004, social networking services (SNS) have appeared one after another, and the mainstream of remote communication has shifted to SNS. Naturally, the convenience of contacting distant friends and family has improved. On the other hand, in a fixed-point survey of daily life conducted by the Hakuhodo Institute of Life and Living (Hakuhodo Institute of Life and Living, n.d.), some findings seem to contradict the improvement in convenience of communication. The percentage of respondents who answered, "I am a person who my friends on various matters often consult" was 41.6% in 1998, compared to 27.8% in the 2020 survey, a drop of nearly 14% in 22 years (Fig. 9.1). These data suggest that the increasing convenience of remote communication has not led to the development of deeper relationships that friends can consult.

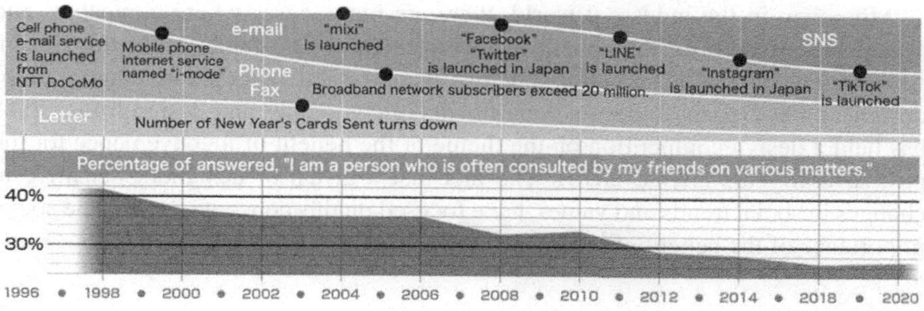

Fig. 9.1 Conceptual diagram showing the evolution of remote communication tools in Japan (Kageyama, 2021)

Thus, we can say that convenience products and services also have some negative aspects. However, the general design philosophy in the modern era implicitly pursues the development of new products that bring about a more convenient and comfortable life than the previous one. In this age of diversified values, pursuing mere convenience does not necessarily lead to essential happiness. We live in an era in which it is even more necessary to search for essential values, to find out what people today are seeking and what things they feel happy about.

9.2 The Benefit of Inconvenience

While the pursuit of convenience is regarded as the basic stance of product and service development, the conflicting concept of "the benefit of inconvenience" is attracting attention. Products and services are designed to achieve some purpose for the user. Taking a camera as an example, we can say that a camera exists to achieve the purpose of capturing people and landscapes in photographs. Thus, convenient functions such as "autofocus" and "auto exposure" can reduce the time and labor required to achieve the user's purpose. On the other hand, the benefit of inconvenience is a concept that is designing more time and labor into the process of achieving the purpose. At first glance, this concept may seem to generate disadvantages for the user. Still, as we have discussed this special issue many times, it is known to create various secondary benefits. In addition, as evidenced by the 3.60 million hits on Instagram when searching for "# Teineina-kurashi (in Japanese)" (reference date: August 2022), there is a movement that values a life that takes time and labor. "Teineina-kurashi" means careful living, a movement to consider the time and labor required to achieve a purpose as an "important factor to feel happiness." For example, switching from cooking rice in an electric rice cooker to cooking rice in an earthenware pot and taking time and labor to cook the rice. This movement suggests there is a social demand for the benefit of inconvenience.

Moreover, in the industrial world, there are many cases where companies and organizations focus on the benefit of inconvenience. The Japan Industrial Design Association (JIDA) is a public interest incorporated association composed of designers who belong to design departments of companies and design firms. JIDA has held a design competition on the theme of the benefit of inconvenience for the past several years (JIDA, 2019). The fact that a group of designers, sensitive to changes in social trends and values, have focused on this concept suggests the social importance of the benefit of inconvenience.

9.3 "Smartification" and "Post Smart"

9.3.1 "Smartification" in Conflict with the Benefit of Inconvenience

The concept of "Smartification" is at odds with the benefit of inconvenience. Smartification is to build intelligent systems that optimize operations according to the situation. This concept is based on optimization using information and communication technology. From the perspective of the relationship between tools and humans, we can say that the initiative of the tool side will be stronger. "Smartification" has been a major development guideline in product design since the IT Revolution occurred in the early 2000s.

9.3.2 "Modernism" and "Post-Modern"

The era goes back to the twentieth century. The Industrial Revolution in the early nineteenth century brought mechanization of product manufacturing and, at the same time, encouraged the separation of design and manufacturing. Products produced by machines became cheap and inferior, and designers emerged to be a new type of professional, capable of designing industrialized products from the perspective of society and human life (JIDA, 2009). Since then, product design has developed through the rise and fall of various styles. In addition, when discussing the history of product design, the connection with war is inseparable. From the perspective of product design, the two world wars were a major motive for advancing functionalism and rationalization. Pursuing rational and functional products without excessive ornamentation was called "Modernism (Katsui et al., 2017)" and became the basis for product design development in the early and middle 20 centuries.

However, modernism underwent a major change in the mid-1970s. Pop and irrational designs have appeared at odds with the rational and functional, known as "Post-Modern." These phenomena were a reaction against the emotional poverty caused by excessive functionalism and rationalism, and they provided an opportunity to rethink capitalist society, industrialization, and material civilization in the twentieth century (JIDA, 2009).

9.3.3 "Post-Smart" Era

The spread of "Benefit of inconvenience" and "Teineina-kurashi" in response to "Smartification" can be thought of as similar to the composition of "Post-Modern" concerning "Modernism." Under the banner of Smartification, efficient, and convenient products and services are being created one after another. However, history shows that trends in design development have waxed and waned. The author calls the movement that sees time and labor as essential happiness, represented by "Benefit of inconvenience" and "Teineina-kurashi" or "Post-Smart."

9.4 Design Work with the Benefit of Inconvenience

In the midst of drastically changing trends in design, the search for essential value is vital for educational institutions that nurture the next generation of designers. For this reason, the School of Design and Architecture, Nagoya City University, has incorporated the benefit of inconvenience into its product design education program. This paper describes the design work conducted from November 2019 to February 2020 for the JIDA Kansai Student Design Award 2019, a design competition on the benefit of inconvenience. In addition, we will organize the main points when incorporating the benefit of inconvenience into design work by unraveling these design processes.

9.4.1 "Size of Inconvenience" and "Tolerance for Inconvenience"

First of all, we cannot say that the benefit of inconvenience is a concept that arises in all situations, regardless of user attributes, time, location, and so on. To create a product or service with the benefit of inconvenience, it is necessary to organize the conditions that tend to bring about the benefit of inconvenience. Let us consider the example of "coffee ground from beans," which is often picked up as an example of the benefit of inconvenience. We can say that a cup of coffee ground from beans made on a holiday morning is a cup of coffee with a value that cannot be enjoyed with instant coffee. On the other hand, working families are generally busy on weekday mornings and do not have much time to enjoy a cup

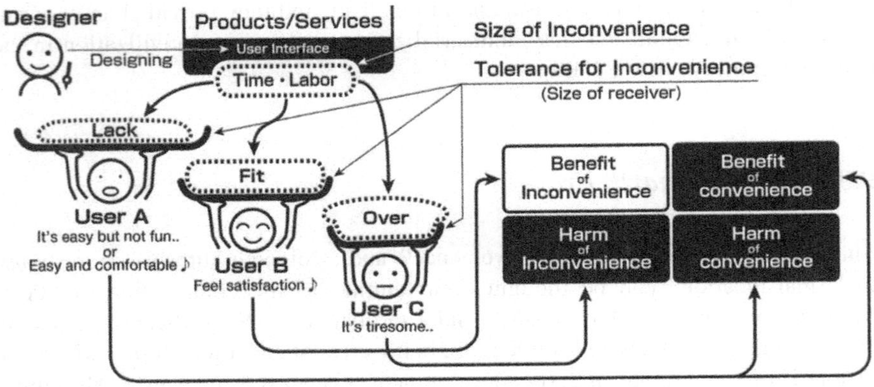

Fig. 9.2 Conceptual diagram showing the relationship between "Size of Inconvenience" and "Tolerance for Inconvenience" (Kageyama, 2021)

of coffee ground from beans. Thus, user attributes, location of use, time, and other factors significantly impact the conditions that cause the benefit of inconvenience. Figure 9.2 is a conceptual diagram showing the relationship between the "Size of Inconvenience" provided by a product or service and the user's "Tolerance for Inconvenience." The author defined the size of inconvenience that can be tolerated, which varies according to user attributes and usage situations, as the "Tolerance for inconvenience." And, when the "size of inconvenience" provided by the product or service and the "Tolerance for inconvenience" on the user side match, the "benefit of inconvenience" which is a value different from convenience and efficiency, is brought about.

9.4.2 *Theme Selection Considering "Tolerance for Inconvenience"*

As mentioned earlier, organizing the relationship between the "Size of inconvenience" and the "Tolerance for Inconvenience" is vital to realizing a design with the benefit of inconvenience. Based on the above, I focused on "travel and pleasure" as a scene that is highly likely to bring the benefit of inconvenience. Travel and pleasure are pastimes that allow people to escape their daily lives. In other words, we can say that the "Tolerance for inconvenience" during travel and pleasure is bigger than in daily life. Thus, I set "Products that bring joy to travel" as the production theme and asked students to consider design proposals.

9.4.3 Design Process to Create the Benefit of Inconveniences

The duration of this design work was 8 weeks, and the process was as follows. In the first week, I explained the assignment and surveyed the theme. In the second week, to analyze the "pleasure of travel," each student created a journey map of their most memorable travel experiences. Journey mapping is a method of UX design (Ando, 2016) that visualizes a series of user experiences across multiple touchpoints, including processes and user behaviors, and emotions. In the third week, I conducted a classroom lecture to understand the concept of benefit of inconvenience. In the fourth week, the students extracted episodes and problems by focusing on the ups and downs of the emotional curve in the journey map they had created. The fifth week was a divergence of ideas, in which ideas diverged from the viewpoint of adding "benefit of inconvenience" to the episodes extracted in the fourth week. From the sixth week onward, the process progressed to convergence of ideas, materialization, and presentation. I will present three representative design proposals produced by students in these practices and their thought processes.

9.4.4 "Matching Pillow" by Chiharu Arima

The "Matching Pillow" is a pillow for a Japanese-style inn that is made by selecting the provided "bag" and "contents" and making it by oneself. Designing a pillow for travelers who often have trouble sleeping because it does not fit their body shape creates the value of "being able to sleep well on a pillow that fits you even when traveling" by designing the time and labor of "making it yourself" (Fig. 9.3) (Kageyama, 2021).

Figure 9.4 is a journey map created by the student of this proposal during the second week (Kageyama, 2021). While analyzing his travel experience, he focused on the episode of "not being able to sleep well because the pillows at the inn did not fit." Next, in the process of diverging ideas to solve the episode's problem, idea divergence was conducted for the benefit of inconvenience. A solution to the problem of "not being able to get a good sleep due to a pillow that doesn't fit" already exists, an example of "multiple pillow selection services" offered by some

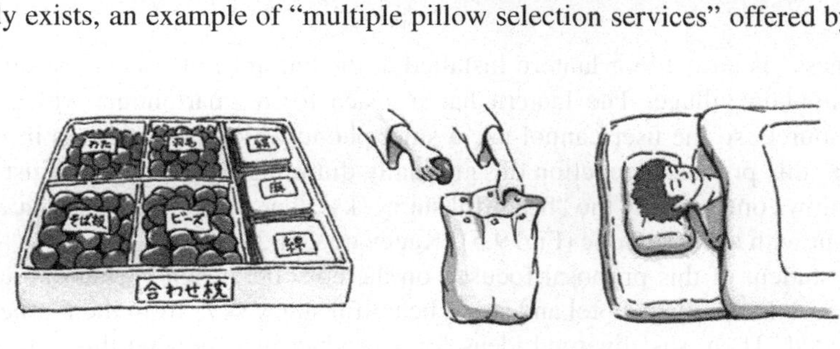

Fig. 9.3 "Matching Pillow"

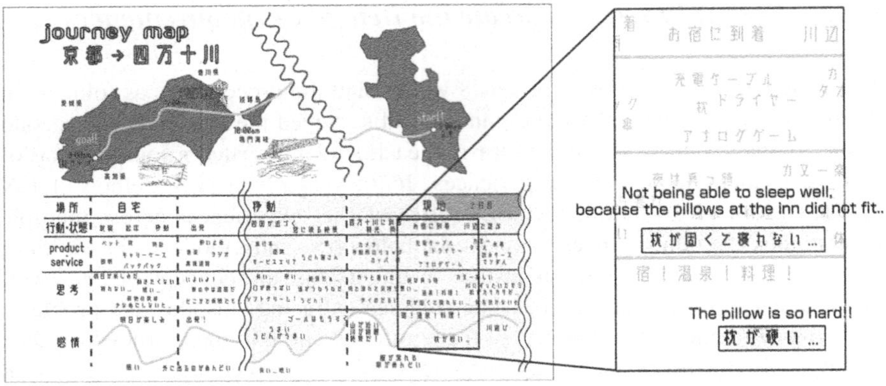

Fig. 9.4 Journey map made by the author of "Matching Pillow"

Fig. 9.5 "Quietness"

business hotels. However, the student has created a new "making a pillow with own hands" style by designing labor and time to achieve the purpose. We can say that this idea could not have been created by the conventional design concept of reducing the time and labor required to achieve a purpose.

9.4.5 "Quietness" by Hiroko Yamamoto

"Quietness" is an outdoor lantern installed at the entrance of a Japanese-style inn in a mountain village. The lantern has a space for a smartphone, which is the power source, so the user cannot use a smartphone while the lantern is in use. In addition, this product's function of "gradually dimming light as eyes adjust to the dark" allows one to enjoy the "beautiful starry sky" that was difficult to notice when going out with a smartphone (Fig. 9.5) (Kageyama, 2021).

The student of this proposal focused on the episode "when I went out at night from a mountain village hotel and saw a beautiful starry sky" from the journey map she created. Then, she diverged ideas for a product that induced this experience. Since the primary purpose of lanterns is to illuminate the surroundings, we can say that the idea of gradually diminishing light is challenging to generate in a conventional design concept. However, from designing for the inconvenience, she created the idea of slowly dimming the lights.

9.4.6 "Gopen "by Kota Aibara

"Gopen" is a coin-operated locker with a pedometer attached to the key. The coin-operated lockers are installed at train stations near tourist attractions, and Gopen users can use them free of charge if they walk a predetermined number of steps. This system induces an action of "walk a lot," and the user's range of activities is more comprehensive. As a result, unexpected discoveries and encounters are made, such as stores in back alleys that you would overlook if you were walking along a standard route and scenery that you will not find in a guidebook. In addition, this design proposal has the potential to become a viable business, as the cost of installing the fee-free lockers could be funded by the surrounding tourist attractions that would benefit from them (Fig. 9.6) (Kageyama, 2021).

In analyzing his travel experiences in a journey map, the student of this proposal focused on "the starting point of a trip is when putting luggage in a coin locker" and attempted to redesign the coin locker. Suppose he follows the conventional design philosophy of reducing time and labor. In that case, he tends to come up with convenient ideas such as "coin-operated lockers that can be reserved via an app" or "coin-operated lockers with web-based information on their location and usage status." However, from the viewpoint of designing time and labor into the experience process, he came up with the idea of "adding a pedometer function to the key, and if the user does not walk a certain number of steps, free service is not available." In addition, this proposal won the Grand Prize at the JIDA Kansai Student Design Award 2019 (JIDA Kansai Bloc, 2019).

Fig. 9.6 "Gopen"

9.5 In Closing

These design proposals were created by students who carried out design work to make the benefit of inconvenience. We can say that each proposal is a design idea that could not have been created by conventional design philosophy, reducing the time and labor required to achieve a purpose. As mentioned at the beginning of this paper, the essence of product design is about designing things and services with an eye to their value to human life and society. It is precise because we live in such an era that the concept of benefit of inconvenience is being embraced by designers whose role is to create happiness for the next generation. In this age, the pursuit of convenience alone will not produce essential happiness.

References

Ando, M. (2016). *User experience design textbook*. Maruzen Publishing.

Hakuhodo Institute of Life and Living (n.d.). Chronological lifestyle survey on the Japanese people 1992–2020, https://seikatsusoken.jp/teiten/answer/820.html, last accessed 2022/9/14.

Japan Institute of Design Promotion (2005). Good Design Award Finder, Good Design Grand Award, https://www.g-mark.org/award/describe/31609, last accessed 2022/9/14.

Japan Institute of Design Promotion (n.d.). About, what is design?, https://www.jidp.or.jp/en/about/firsttime/whatsdesign, last accessed 2022/9/14.

JIDA. (2009). *Product design—For everyone involved in product development (in Japanese)*. Works Corporation.

JIDA Kansai Bloc, Application Information of JIDA Kansai Student Design Award 2019 (in Japanese), http://www.kansai.jida.or.jp/info/95.html, last accessed 2022/9/14.

JIDA Kansai Bloc, Restart information of JIDA Kansai Student Design Award 2019 (in Japanese), http://www.kansai.jida.or.jp/info/101, last accessed 2022/9/14.

Kageyama, T. (2021). Product design in "post-smart era". *Journal of the Society of Instrument and Control Engineers, 60*(12), 875–879.

Katsui, M., Tanaka, I., & Mukai, S. (2017). *Heibonsya dictionary of Today's design*. Heibonsya.

Ministry of Internal Affairs and Communications (n.d.). How ICT and the digital economy have evolved (in Japanese), https://www.soumu.go.jp/johotsusintokei/whitepaper/ja/r01/pdf/index.html, last accessed 2022/9/14.

Tomoaki Kageyama is a lecturer at the Nagoya City University Graduate School of Design and Architecture. He specializes in product design. Before working at the university, he worked at a design firm, where he designed vehicles, furniture, lifestyle, and industrial equipment. Major awards include Eco Japan Cup 2011 Second Prize, 13th Fuji-no-Kuni Universal Design Award Grand Prize, and CERAMIC LIFE DESIGN AWARD 2018 Honorable Mention. Master of Design from Aichi Public University of Fine Arts and Music.

The manufacturer's authorised representative in the EU is Springer
Nature Customer Service Centre GmbH, Europaplatz 3, 69115 Heidelberg,
Germany. If you have any concerns regarding our products, please
contact ProductSafety@springernature.com

Printed and bound by CPI Group (UK) Ltd, Croydon, CR0 4YY
30/09/2026
02225259-0001